Recommended Residential Construction for Coastal Areas

Building on Strong and Safe Foundations

FEMA P-550, Second Edition / December 2009

About the Cover

On August 29, 2005, Hurricane Katrina struck the Gulf Coast with recordbreaking storm surge that destroyed foundations and devastated homes from Louisiana east to Alabama. Katrina was so destructive that engineers assessing the carnage no longer looked for "success stories" (i.e., homes that were only moderately damaged), but rather searched for "survivor" homes that, while extensively damaged, still bore a slight resemblance to a residential building. Hurricane Katrina proved that, without strong foundations, homes on the coast can and will be destroyed.

RECOMMENDED RESIDENTIAL CONSTRUCTION FOR COASTAL AREAS

Building on Strong and Safe Foundations

Preface

Since the publication of the First Edition of FEMA 550 in July 2006, several advances have been made in nationally adopted codes and standards. Two editions of the International Residential Code® (the 2006 IRC® and the 2009 IRC) and the International Building Code® (the 2006 IBC® and the 2009 IBC) have been published and the long awaited International Code Council (ICC) 600 *Standard for Residential Construction in High Wind Areas* (a successor to the legacy standard SSTD-10) has been issued.

To keep pace with developing codes and standards and to improve its guidance, FEMA is issuing this Second Edition of FEMA 550. In addition to being renamed to more accurately reflect its applicability, the Second Edition of FEMA 550 contains a new foundation style Case H, which incorporates an elevated concrete beam for improved structural efficiency. The Second Edition of FEMA 550 has also been updated for consistency with the 2006 and 2009 editions of the IRC and IBC, and the 2005 Edition of ASCE 7 *Minimum Design Loads for Buildings and Other Structures*.

RECOMMENDED RESIDENTIAL CONSTRUCTION FOR COASTAL AREAS

Building on Strong and Safe Foundations

Acknowledgments

FEMA would like to thank the following individuals who provided information, data, review, and guidance in developing the Second Edition of this publication.

FEMA

John Ingargiola
FEMA Headquarters

Consultants

Dave Conrad
PBSJ

Deb Daly
Greenhorne & O'Mara, Inc.

Julie Liptak
Greenhorne & O'Mara, Inc.

David K. Low, PE
DK Low & Associates, LLC

Kelly Park
Greenhorne & O'Mara, Inc.

Scott Sundberg
Category X Coastal Consulting

Scott Tezak
URS Corporation

Jimmy Yeung, PhD, PE
Greenhorne & O'Mara, Inc.

In addition, FEMA would like to acknowledge the members of the project team for the First Edition of the publication. (Note: all affiliations were current as of July 2006.)

ACKNOWLEDGMENTS

Principal Authors

Bill Coulbourne, PE
URS Corporation

Matt Haupt, PE
URS Corporation

Scott Sundberg, PE
URS Corporation

David K. Low, PE
DK Low & Associates, LLC

Jimmy Yeung, PhD, PE
Greenhorne & O'Mara, Inc.

John Squerciati, PE
Dewberry & Davis, LLC

Contributors and Reviewers

John Ingargiola
FEMA Headquarters

Shabbar Saifee
FEMA, Region IV

Dan Powell
FEMA, Region IV

Alan Springett
FEMA, Region IV

Keith Turi
FEMA Headquarters

Christopher Hudson
FEMA Headquarters

Christopher P. Jones, PE

Dan Deegan, PE, CFM
PBSJ

Ken Ford
National Association of Homebuilders (NAHB)

Mike Hornbeck
Gulf Construction Company, Inc.

David Kriebel, PhD, PE
U.S. Naval Academy

Jim Puglisi
Dewberry & Davis, LLC

John Ruble
Bayou Plantation Homes

Bob Speight, PE
URS Corporation

Deb Daly
Greenhorne & O'Mara, Inc.

Julie Liptak
Greenhorne & O'Mara, Inc.

Wanda Rizer
Design4Impact

Naomi Chang Zajic
Greenhorne & O'Mara, Inc.

RECOMMENDED RESIDENTIAL CONSTRUCTION
FOR COASTAL AREAS

Building on Strong and Safe Foundations

Table of Contents

Preface iii

Acknowledgments v

Introduction xv

Chapter 1. Types of Hazards 1-1
1.1   High Winds 1-1
1.2   Storm Surge 1-5
1.3   Flood Effects 1-5
      1.3.1   Hydrostatic Forces 1-7
      1.3.2   Hydrodynamic Forces 1-7
      1.3.3   Waves 1-8
      1.3.4   Floodborne Debris 1-10
      1.3.5   Erosion and Scour 1-10

Chapter 2. Foundations 2-1
2.1   Foundation Design Criteria 2-1

TABLE OF CONTENTS

2.2	Foundation Design in Coastal Areas		2-2
2.3	Foundation Styles in Coastal Areas		2-4
	2.3.1	Open Foundations	2-5
		2.3.1.1 Piles	2-5
		2.3.1.2 Piers	2-7
	2.3.2	Closed Foundations	2-8
		2.3.2.1 Perimeter Walls	2-8
		2.3.2.2 Slab-on-Grade	2-10
2.4	Introduction to Foundation Design and Construction		2-11
	2.4.1	Site Characterization	2-11
	2.4.2	Types of Foundation Construction	2-12
		2.4.2.1 Piles	2-12
		2.4.2.2 Diagonal Bracing of Piles	2-13
		2.4.2.3 Knee Bracing of Piles	2-14
		2.4.2.4 Wood-Pile-to-Wood-Girder Connections	2-15
		2.4.2.5 Grade Beams in Pile/Column Foundations	2-15

Chapter 3. Foundation Design Loads 3-1

3.1	Wind Loads		3-2
3.2	Flood Loads		3-2
	3.2.1	Design Flood and DFE	3-2
	3.2.2	Design Stillwater Flood Depth (d_s)	3-4
	3.2.3	Design Wave Height (H_b)	3-5
	3.2.4	Design Flood Velocity (V)	3-5
3.3	Hydrostatic Loads		3-6
3.4	Wave Loads		3-7
	3.4.1	Breaking Wave Loads on Vertical Piles	3-8
	3.4.2	Breaking Wave Loads on Vertical Walls	3-8
3.5	Hydrodynamic Loads		3-9
3.6	Debris Impact Loads		3-12

3.7	Erosion and Localized Scour	3-14
	3.7.1 Localized Scour Around Vertical Piles	3-15
	3.7.2 Localized Scour Around Vertical Walls and Enclosures	3-19
3.8	Flood Load Combinations	3-19

Chapter 4. Overview of Recommended Foundation Types and Construction for Coastal Areas 4-1

4.1	Critical Factors Affecting Foundation Design	4-2
	4.1.1 Wind Speed	4-2
	4.1.2 Elevation	4-3
	4.1.3 Construction Materials	4-4
	4.1.3.1 Masonry Foundation Construction	4-4
	4.1.3.2 Concrete Foundation Construction	4-4
	4.1.3.3 Field Preservative Treatment for Wood Members	4-5
	4.1.4 Foundation Design Loads	4-5
	4.1.5 Foundation Design Loads and Analyses	4-8
4.2	Recommended Foundation Types for Coastal Areas	4-14
	4.2.1 Open/Deep Foundation: Timber Pile (Case A)	4-15
	4.2.2 Open/Deep Foundation: Steel Pipe Pile with Concrete Column and Grade Beam (Case B)	4-17
	4.2.3 Open/Deep Foundation: Timber Pile with Concrete Column and Grade Beam (Case C)	4-17
	4.2.4 Open/Deep Foundation: Timber Pile with Concrete Grade and Elevated Beams and Concrete Columns (Case H)	4-19
	4.2.5 Open/Shallow Foundation: Concrete Column and Grade Beam with Slabs (Cases D and G)	4-21
	4.2.6 Closed/Shallow Foundation: Reinforced Masonry – Crawlspace (Case E)	4-21
	4.2.7 Closed/Shallow Foundation: Reinforced Masonry – Stem Wall (Case F)	4-23

Chapter 5. Foundation Selection 5-1

5.1	Foundation Design Types	5-1
5.2	Foundation Design Considerations	5-2
5.3	Cost Estimating	5-4

| 5.4 | How to Use This Manual | 5-4 |
| 5.5 | Design Examples | 5-7 |

Appendices

Appendix A Foundation Designs

Appendix B Pattern Book Design

Appendix C Assumptions Used in Design

Appendix D Foundation Analysis and Design Examples

Appendix E Cost Estimating

Appendix F Pertinent Coastal Construction Information

Appendix G FEMA Publications and Additional References

Appendix H Glossary

Appendix I Abbreviations and Acronyms

Tables

Chapter 2

| Table 2-1. | Foundation Type Dependent on Coastal Area | 2-5 |

Chapter 3

Table 3-1.	Building Category and Corresponding Dynamic Pressure Coefficient (C_p)	3-9
Table 3-2.	Drag Coefficient Based on Width to Depth Ratio	3-11
Table 3-3.	Example Foundation Adequacy Calculations for a Two-Story Home Supported on Square Timber Piles	3-17
Table 3-4.	Local Scour Depth as a Function of Soil Type	3-19
Table 3-5.	Selection of Flood Load Combinations for Design	3-21

Chapter 4

| Table 4-1a. | Design Perimeter Wall Reactions (lb/lf) for One-Story Elevated Homes | 4-7 |
| Table 4-1b. | Design Perimeter Wall Reactions (lb/lf) for Two-Story Elevated Homes | 4-7 |

Table 4-2.	Flood Forces (in pounds) on an 18-Inch Square Column	4-7
Table 4-3.	Wind Reactions Used to Develop Case H Foundations	4-8
Table 4-4.	Design Moments (K-ft), Axial Loads (in kips), and Shears (in kips) for 10-Foot Tall 3-Bay Foundations	4-10
Table 4-5.	Design Moments (K-ft), Axial Loads (in kips), and Shears (in kips) for 15-Foot Tall 3-Bay Foundations	4-11
Table 4-6.	Design Moments (K-ft), Axial Loads (in kips), and Shears (in kips) for 10-Foot Tall 6-Bay Foundations	4-12
Table 4-7.	Design Moments (K-ft), Axial Loads (in kips), and Shears (in kips) for 15-Foot Tall 6-Bay Foundations	4-12
Table 4-8.	Design Moments (K-ft), Axial Loads (in kips), and Shears (in kips) for 10-Foot Tall 9-Bay Foundations	4-13
Table 4-9.	Design Moments (K-ft), Axial Loads (in kips), and Shears (in kips) for 15-Foot Tall 9-Bay Foundations	4-14
Table 4-10.	Recommended Foundation Types Based on Zone	4-15

Chapter 5

Table 5-1a.	Foundation Design Cases for One-Story Homes Based on Height of Elevation and Wind Velocity	5-10
Table 5-1b.	Foundation Design Cases for Two-Story Homes Based on Height of Elevation and Wind Velocity	5-11

Figures

Introduction

Figure 1.	Damage to residential properties as a result of Hurricane Katrina's winds and storm surge. Note the building that was knocked off its foundation	xvi
Figure 2.	Schematic range of home dimensions and roof pitches used as the basis for the foundation designs presented in this manual.	xviii

Chapter 1

Figure 1-1.	Wind damage to roof structure and gable end wall from Hurricane Katrina (2005) (Pass Christian, Mississippi).	1-2
Figure 1-2.	Saffir-Simpson Scale.	1-3

TABLE OF CONTENTS

Figure 1-3.   Wind speeds (in mph) for the entire U.S. 1-4

Figure 1-4.   Graphical depiction of a hurricane moving ashore. In this example, a 15-foot surge added to the normal 2-foot tide creates a total storm tide of 17 feet. 1-5

Figure 1-5.   Storm tide and waves from Hurricane Dennis on July 10, 2005, near Panacea, Florida. 1-6

Figure 1-6.   Comparison of storm surge levels along the shorelines of the Gulf Coast for Category 1, 3, and 5 storms. 1-6

Figure 1-7.   Building floated off of its foundation (Plaquemines Parish, Louisiana). 1-7

Figure 1-8.   Aerial view of damage to one of the levees caused by Hurricane Katrina. 1-8

Figure 1-9.   During Hurricane Opal (1995), this house was in an area of channeled flow between large buildings. As a result, the house was undermined and washed into the bay behind a barrier island. 1-8

Figure 1-10. Storm waves breaking against a seawall in front of a coastal residence at Stinson Beach, California. 1-9

Figure 1-11. Storm surge and waves overtopping a coastal barrier island in Alabama (Hurricane Frederic, 1979). 1-9

Figure 1-12. Pier piles were carried over 2 miles by the storm surge and waves of Hurricane Opal (1995) before coming to rest in Pensacola Beach, Florida. .. 1-10

Figure 1-13. Extreme case of localized scour undermining a slab-on-grade house in Topsail Island, North Carolina, after Hurricane Fran (1996) 1-11

Chapter 2

Figure 2-1.   Recommended open foundation practice for buildings in A zones, Coastal A zones, and V zones. 2-3

Figure 2-2   Slab-on-grade foundation failure due to erosion and scour undermining from Hurricane Dennis, 2005 (Navarre Beach, Florida). 2-3

Figure 2-3.   Compression strut at base of a wood pile. Struts provide some lateral support for the pile, but very little resistance to rotation. 2-6

Figure 2-4.   Near collapse due to insufficient pile embedment (Dauphin Island, Alabama). 2-6

Figure 2-5.   Successful pile foundation following Hurricane Katrina (Dauphin Island, Alabama). 2-6

TABLE OF CONTENTS

Figure 2-6.	Column connection failure (Belle Fontaine Point, Jackson County, Mississippi).	2-7
Figure 2-7.	Performance comparison of pier foundations. Piers on discrete footings failed while piers on more substantial footings survived (Pass Christian, Mississippi).	2-8
Figure 2-8.	Isometric view of an open foundation with grade beam.	2-9
Figure 2-9.	Isometric view of a closed foundation with crawlspace.	2-10
Figure 2-10.	Pile installation methods.	2-12
Figure 2-11.	Diagonal bracing schematic.	2-14

Chapter 3

Figure 3-1.	Wind speeds (in mph) for the entire U.S.	3-3
Figure 3-2.	Parameters that determine or are affected by flood depth.	3-4
Figure 3-3.	Normally incident breaking wave pressures against a vertical wall (space behind vertical wall is dry)	3-10
Figure 3-4.	Normally incident breaking wave pressures against a vertical wall (stillwater level equal on both sides of wall).	3-10
Figure 3-5.	Distinguishing between coastal erosion and scour. A building may be subject to either or both, depending on the building location, soil characteristics, and flood conditions.	3-14
Figure 3-6.	Scour at vertical foundation member stopped by underlying scour-resistant stratum.	3-16

Chapter 4

Figure 4-1.	The BFE, freeboard, erosion, and ground elevation determine the foundation height required.	4-3
Figure 4-2.	Design loads acting on a column.	4-6
Figure 4-3.	Shear panel reactions for the 3- and 6-bay models. Reactions for the 9-bay model were similar to those of the 6-bay.	4-10
Figure 4-4.	Profile of Case A foundation type.	4-16
Figure 4-5.	Profile of Case B foundation type.	4-18
Figure 4-6.	Profile of Case C foundation type.	4-19

TABLE OF CONTENTS

Figure 4-7. Profile of Case H foundation type. 4-20

Figure 4-8. Profile of Case G foundation type. 4-22

Figure 4-9. Profile of Case D foundation type. 4-23

Figure 4-10. Profile of Case F foundation type. 4-24

Figure 4-11. Profile of Case E foundation type. 4-24

Chapter 5

Figure 5-1. Schematic of a basic module and two footprints. 5-3

Figure 5-2. Foundation selection decision tree 5-8

Figure 5-3. "T" shaped modular design 5-12

Figure 5-4. "L" shaped modular design 5-12

Figure 5-5. "Z" shaped modular design 5-13

RECOMMENDED RESIDENTIAL CONSTRUCTION
FOR COASTAL AREAS

Building on Strong and Safe Foundations

Introduction

The purpose of this design manual is to provide recommended foundation designs and guidance for rebuilding homes destroyed by hurricanes in coastal areas. In addition, the manual is intended to provide guidance in designing and building safer and less vulnerable homes to reduce the risk to life and property.

Past storms such as Hurricanes Andrew, Hugo, Charley, Katrina, and Rita, and recent events such as Hurricane Ike continue to show the vulnerability of our "built environment" (Figure 1). While good design and construction cannot totally eliminate risk, every storm has shown that sound design and construction can significantly reduce the risk to life and damage to property. With that in mind, the Federal Emergency Management Agency (FEMA) has developed this manual to help the community of homebuilders, contractors, and local engineering professionals in rebuilding homes destroyed by hurricanes, and designing and building safer and less vulnerable new homes.

INTRODUCTION

Figure 1.
Damage to residential properties as a result of Hurricane Katrina's winds and storm surge. Note the building that was knocked off its foundation (circled).

SOURCE: HURRICANE KATRINA MAT PHOTO

Intent of the Manual

The intent of the manual is to provide homebuilders, contractors, and engineering professionals with a series of recommended foundation designs that will help create safer and stronger buildings in coastal areas. The designs are intended to help support rebuilding efforts after coastal areas have been damaged by floods, high winds, or other natural hazards.

The foundations may differ somewhat from traditional construction techniques; however, they represent what are considered to be some of the better approaches to constructing strong and safe foundations in hazardous coastal areas. The objectives used to guide the development of this manual are:

- To provide residential foundation designs that will require minimal engineering oversight
- To provide foundation designs that are flexible enough to accommodate many of the homes identified in *A Pattern Book for Gulf Coast Neighborhoods* prepared for the Mississippi Governor's Rebuilding Commission on Recovery, Rebuilding, and Renewal (see Appendix B)
- To utilize "model" layouts so that many homes can be constructed without significant additional engineering efforts

The focus of this document is on the foundations of residential buildings. The assumption is that those who are designing and building new homes will be responsible for ensuring that the building itself is designed according to the latest building code (International Building Code® [IBC®], International Residential Code® [IRC®], and FEMA guidance) and any local requirements. The user of this manual is directed to other publications that also address disaster-resistant construction (see Appendix G).

INTRODUCTION

Although the foundation designs are geared to the coastal environment subject to storm surge, waves, floating debris, and high winds, several are suitable for supporting homes on sites protected by levees and floodwalls or in riverine areas subjected to high-velocity flows. Design professionals can be contacted to ensure the foundation designs provided in this manual are suitable for specific sites.

This manual contains closed foundation designs for elevating homes up to 8 feet above ground level and open foundation designs for elevating homes up to 15 feet above ground level. These upper limits are a function of constructability limitations and overturning and stability issues for more elevated foundations. Eight-foot tall foundations are a practical upper limit for 8-inch thick reinforced concrete masonry unit (CMU) walls exposed to flood forces anticipated in non-coastal A zones. The upper limit of 15 feet for open foundations was established by estimating the amount a home needs to be elevated to achieve the 2005 Advisory Base Flood Elevations (ABFEs) as determined in response to Hurricanes Katrina and Rita. The ABFEs published in the Hurricane Recovery Maps were compared to local topographic maps for the Gulf Coast. The comparison revealed that providing foundation designs up to 15 feet tall would allow over 80 percent of the homes damaged by Hurricane Katrina to be protected from flood events reflected in the ABFEs and in the Hurricane Recovery Maps. Many homes can be elevated to the BFE on foundations that are 4 feet tall or less.

CAUTION: Although sites inside levees are not exposed to wave loads, sites immediately adjacent to floodwalls and levees can be exposed to extremely high flood velocities and scour if a breach occurs. Design professionals should be consulted before using these foundation designs on sites close to floodwalls or levees to determine if they are appropriate.

This updated edition of FEMA 550 introduces the Case H foundation, which is an open/deep foundation developed for use in coastal high hazard areas (V zones). It is also appropriate to use the Case H foundation in Coastal A and non-coastal A zones. Case H foundations incorporate elevated reinforced concrete beams that provide three important benefits. One, the elevated beams work in conjunction with the reinforced concrete columns and grade beams to produce a structural frame that is more efficient at resisting lateral loads than the grade beams and cantilevered columns used in other FEMA 550 open foundations. The increased efficiency allows foundations to be constructed with smaller columns that are less exposed to flood forces.

The second benefit is that the elevated reinforced concrete beams provide a continuous foundation that can support many homes constructed to prescriptive designs from codes and standards such as the IRC, the American Forest and Paper Association's (AF&PA's) *Wood Frame Construction Manual for One- and Two-Family Dwellings (WFCM)*, and the International Code Council's (ICC's) *Standard for Residential Construction in High Wind Regions* (ICC-600).

The third benefit that Case H foundations provide is the ability to support relatively narrow (14-foot wide) homes. It is anticipated that Case H foundations can be used for several styles of modular homes.

INTRODUCTION

Using the Manual

The following information is needed to use this manual:

- Design wind speed and the Design Flood Elevation (DFE) at the site
- The flood zone(s) at the site
- Building layout
- Topographic elevation of existing building site
- Soil conditions for the site. Soil condition assumptions used in the load calculations are intentionally conservative. Users are encouraged to determine soil conditions at the site to potentially improve the cost-effectiveness of the design.

Most of the information can be obtained from the local building official or floodplain manager.

This document is not intended to supplant involvement from local design professionals. While the designs included can be used without modification (provided that the home to be elevated falls within the design criteria), consulting with local engineers should be considered. Local engineers may assist with the following:

- Incorporating local site conditions into the design
- Addressing and supporting unique features of the home
- Confirming the suitability of the designs for a specific home on a specific site
- Allowing use of value engineering to produce a more efficient design

These "prescriptive" designs have been developed to support homes with a range of dimensions, weights, and roof pitches. Figure 2 schematically shows the diverse range of dimensions

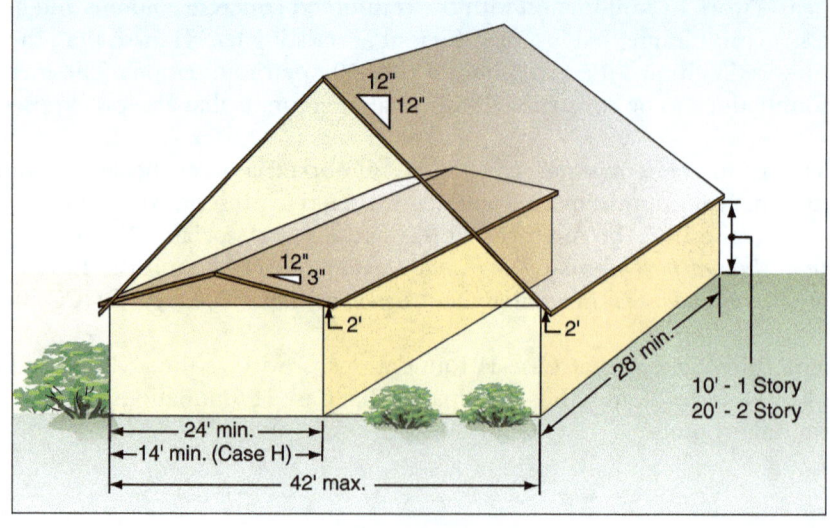

Figure 2. Schematic range of home dimensions and roof pitches used as the basis for the foundation designs presented in this manual.

and roof pitches. Appendix C contains a complete list of criteria and assumptions used in these designs.

This manual concentrates on foundations that resist the extreme hurricane wind and flooding conditions found in many coastal areas. For successful, natural hazard-resistant installations, both the foundation and the home it supports must be properly designed and constructed to take all loads on the structure into the ground through the foundation. Designing and constructing the home to meet all the requirements of the IBC or the IRC is the minimum action necessary to producing hazard-resistant homes. However, any model code must contain minimum requirements. FEMA supports the use of best practice approaches for improved resistance to natural hazards.

The foundations presented in this manual have been designed to resist the flood, wind, and gravity loads specified in Appendix C and load combinations specified in the American Society of Civil Engineers' (ASCE's) *Minimum Design Loads for Buildings and Other Structures* (ASCE 7-05). Although the designs provide significant resistance to gravity, lateral, and uplift loads, they have not been specifically designed for seismic events. In coastal areas where seismic risks exist, design professionals should confirm that the foundation designs presented herein are adequate to resist seismic loads on site.

To gain the benefits of a "best practices" approach, readers are directed to publications such as FEMA 499, *Home Builder's Guide to Coastal Construction Technical Fact Sheet Series*, and FEMA 55, *Coastal Construction Manual*. A more complete list of available publications is contained in Appendix G.

Organization of the Manual

There are five chapters and nine appendices in this manual. The intent is to cover the essential information in the chapters and provide all the details in the appendices. Chapter 1 provides a description of the different types of hazards that must be considered in the design of a residential building foundation in a coastal area. The primary issues related to designing foundations for residential buildings are described in Chapter 2. Chapter 3 provides guidance on how to determine the magnitude of the loads placed on a building by a particular natural hazard event or a combination of events. The different foundation types and methods of construction foundation for a residential building are discussed in Chapter 4. Chapter 5 and Appendix A present foundation designs to assist the homebuilders, contractors, and local engineering professionals in developing safe and strong foundations.

In addition to Chapters 1 through 5 and Appendix A, the following appendices are presented herein:

- Appendix B presents examples of how the foundation designs in this manual can be used with some of the homes in the publication *A Pattern Book for Gulf Coast Neighborhoods*.

- Appendix C provides a list of assumptions used in developing the foundation design presented in this manual.

INTRODUCTION

- Appendix D presents detailed calculations on how to design the foundation of residential buildings. Two examples, one for open foundations and the other for closed foundations, are included.

- Appendix E provides cost information that the homebuilders can use to estimate the cost of installing the foundation systems proposed in this manual.

- Appendix F includes fact sheets contained in FEMA 499 (*Home Builder's Guide to Coastal Construction Technical Fact Sheet Series*) and a recovery advisory contained in FEMA P-757 (*Hurricane Ike in Texas and Louisiana: Mitigation Assessment Team Report, Building Performance Observations, Recommendations, and Technical Guidance*) that are pertinent to construction in coastal areas.

- Appendix G presents a list of references and other FEMA publications that can be of assistance to the users of this manual.

- Appendix H contains a glossary of terms used in this manual.

- Appendix I defines abbreviations and acronyms used in this manual.

Limitations of the Manual

This manual has been provided to assist in reconstruction efforts after coastal areas have been damaged by floods, high winds, or other natural hazards. Builders, architects, or engineers using this manual assume responsibility for the resulting designs and their performance during a natural hazard event.

The foundation designs and analyses presented in this manual were based on ASCE 7-02 and the 2003 version of the IRC. While FEMA 550 was being developed, ASCE released its 2005 edition of ASCE 7 (ASCE 7-05) and the ICC issued their 2006 editions of the IBC and IRC. The ASCE 7 revisions did not affect the load calculations controlling the designs and there were no substantive flood provision changes to the IRC that affect foundation designs in coastal areas.

This Second Edition of FEMA 550 is consistent with the 2009 editions of the IBC and IRC.

RECOMMENDED RESIDENTIAL CONSTRUCTION FOR COASTAL AREAS

Building on Strong and Safe Foundations

1. Types of Hazards

This chapter discusses the following types of hazards that must be considered in the design of a residential building foundation for coastal areas: high winds, storm surge, and associated flood effects, including hydrostatic forces, hydrodynamic forces, waves, floodborne debris, and erosion and scour.

1.1 High Winds

Hurricanes and typhoons are the basis for design wind speeds for many portions of the U.S. and its territories. High winds during a hurricane can create extreme positive and negative forces on a building; the net result is that wind forces simultaneously try to push over the building and lift it off its foundation. If the foundation is not strong enough to resist these forces, the home may slide, overturn, collapse, or incur substantial damage (Figure 1-1).

1 TYPES OF HAZARDS

The most current design wind speeds are provided by the American Society of Civil Engineers (ASCE) document *Minimum Design Loads for Buildings and Other Structures* (ASCE 7). ASCE 7 is typically updated every 3 years. The 2002 edition (ASCE 7-02) is referenced by the following model building codes: the 2009 editions of the International Building Code® (IBC®) and the International Residential Code® (IRC®).

NOTE: Hurricanes are classified into five categories according to the Saffir-Simpson Scale, which uses wind speed and central pressure as the principal parameters to categorize storm damage potential. Hurricanes can range from Category 1 to the devastating Category 5 (Figure 1-2).

Hurricanes can produce storm surge that is higher or lower than what the wind speed at landfall would predict. While Hurricane Katrina's storm surge was roughly that of a Category 5, its winds at landfall were only a Category 3. A hurricane that is a Category 3 or above is generally considered a major hurricane.

Design wind speeds given by ASCE 7 are 3-second gust speeds, not the sustained wind speeds associated with the Saffir-Simpson hurricane classification scale (Figure 1-2). Figure 1-3 shows the design wind speeds for portions of the Gulf Coast region based on 3-second gusts (measured at 33 feet above the ground in Exposure C).

Figure 1-1.
Wind damage to roof structure and gable end wall from Hurricane Katrina (2005) (Pass Christian, Mississippi).

SOURCE: HURRICANE KATRINA MAT PHOTO

TYPES OF HAZARDS 1

Saffir-Simpson Scale (Category/Damage)

Category 1 Hurricane – Winds 74 to 95 mph, sustained (91 to 116 mph, 3-second gust)

No real damage to buildings. Damage to unanchored mobile homes. Some damage to poorly constructed signs. Also, some coastal flooding and minor pier damage. Examples: Hurricanes Irene (1999) and Allison (1995).

Category 2 Hurricane – Winds 96 to 110 mph, sustained (117 to 140 mph, 3-second gust)

Some damage to building roofs, doors, and windows. Considerable damage to mobile homes. Flooding damages piers, and small crafts in unprotected moorings may break. Some trees blown down. Examples: Hurricanes Bonnie (1998), Georges (FL and LA, 1998), and Gloria (1985).

Category 3 Hurricane – Winds 111 to 130 mph, sustained (141 to 165 mph, 3-second gust)

Some structural damage to small residences and utility buildings. Large trees blown down. Mobile homes and poorly built signs destroyed. Flooding near the coast destroys smaller structures with larger structures damaged by floating debris. Terrain may be flooded far inland. Examples: Hurricanes Keith (2000), Fran (1996), Opal (1995), Alicia (1983), Betsy (1965), and Katrina at landfall (2005).

Category 4 Hurricane – Winds 131 to 155 mph, sustained (166 to 195 mph, 3-second gust)

More extensive curtainwall failures with some complete roof structure failure on small residences. Major erosion of beach areas. Terrain may be flooded far inland. Examples: Hurricanes Hugo (1989), Donna (1960), and Charley (2004).

Category 5 Hurricane – Winds greater than 155 mph, sustained (195 mph and greater, 3-second gust)

Complete roof failure on many residences and industrial buildings. Some complete building failures with small utility buildings blown over or away. Flooding causes major damage to lower floors of all structures near the shoreline. Massive evacuation of residential areas may be required. Examples: Hurricanes Andrew (1992), Camille (1969), and the unnamed Labor Day storm (1935).

Note: Saffir-Simpson wind speeds (sustained 1-minute) were converted to 3-second gust wind speed utilizing the Durst Curve contained in ASCE 7-05, Figure C6-2.

Figure 1-2. Saffir-Simpson Scale.
SOURCE: *HURRICANE KATRINA IN THE GULF COAST* (FEMA 549)

1 TYPES OF HAZARDS

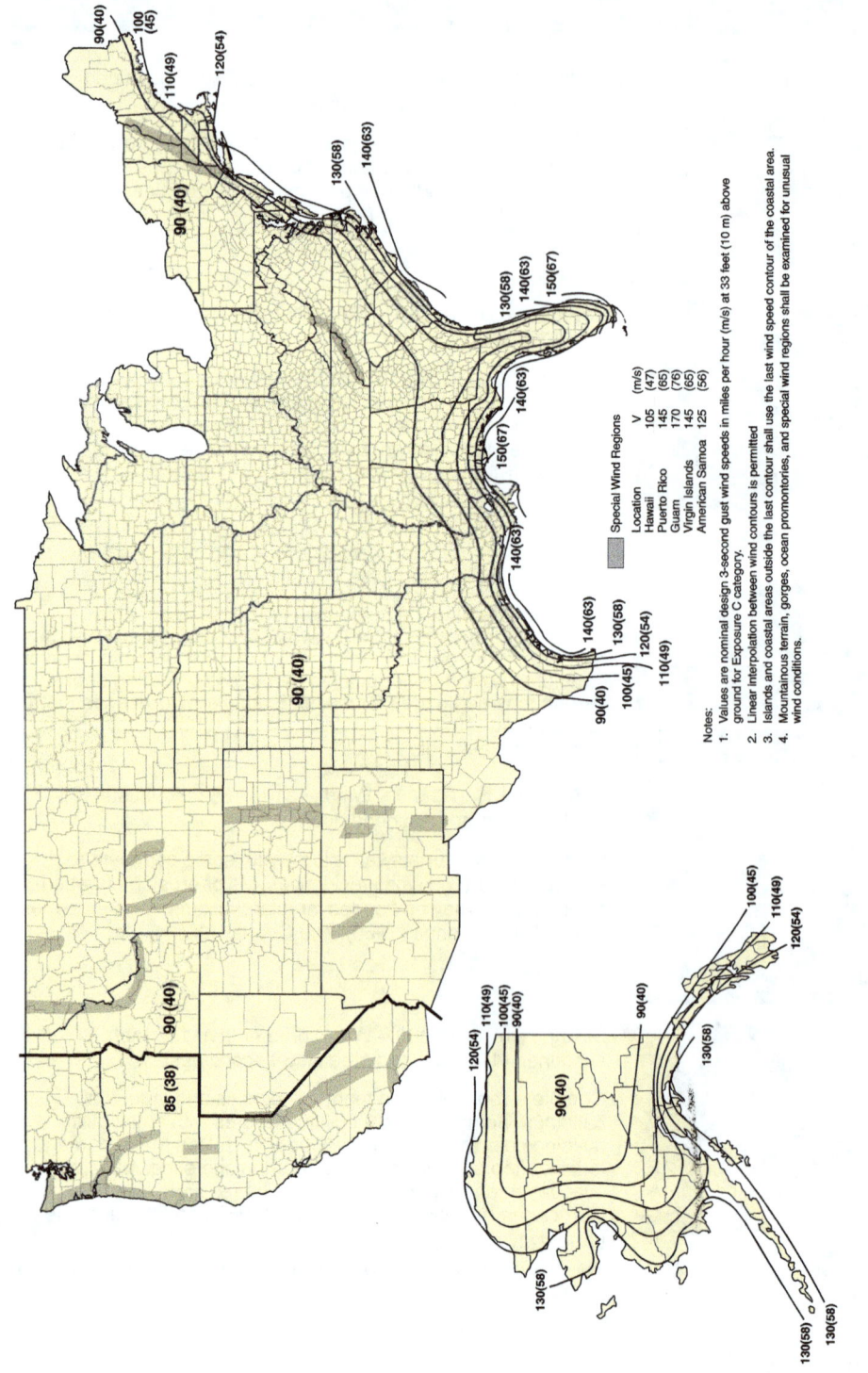

Figure 1-3. Wind speeds (in mph) for the entire U.S.
SOURCE: ASCE 7-05

1.2 Storm Surge

Storm surge is water that is pushed toward the shore by the combined force of the lower barometric pressure and the wind-driven waves advancing to the shoreline. This advancing surge combines with the normal tides to create the hurricane storm tide, which in many areas can increase the sea level by as much as 20 to 30 feet. Figure 1-4 is a graphical depiction of how wind-driven waves are superimposed on the storm tide. This rise in water level can cause severe flooding in coastal areas, particularly when the storm tide coincides with high tides (Figure 1-5). Because much of the United States' densely populated coastlines lie less than 20 feet above sea level, the danger from storm surge is great. This is particularly true along the Gulf of Mexico where the shape and bathymetry of the Gulf contribute to storm surge levels that can exceed most other areas in the U.S. (Figure 1-6).

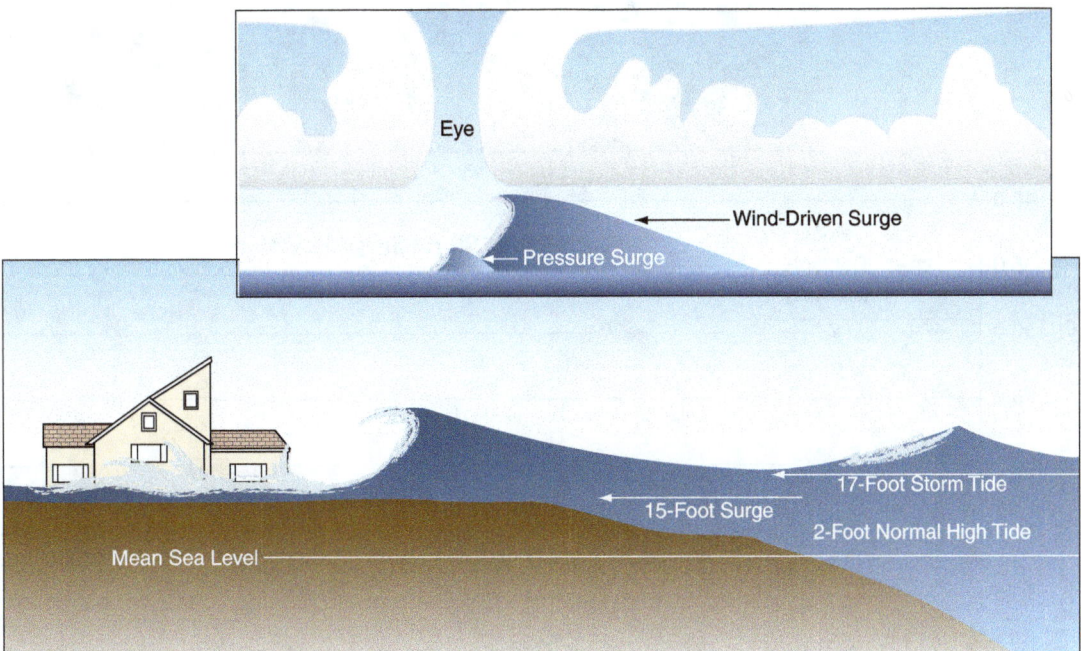

Figure 1-4.
Graphical depiction of a hurricane moving ashore. In this example, a 15-foot surge added to the normal 2-foot tide creates a total storm tide of 17 feet.

1.3 Flood Effects

Although coastal flooding can originate from a number of sources, hurricanes and weaker tropical storms not categorized as hurricanes are the primary cause of flooding (Figure 1-2). The flooding can lead to a variety of impacts on coastal buildings and their foundations: hydrostatic forces, hydrodynamic forces, waves, floodborne debris forces, and erosion and scour.

1 TYPES OF HAZARDS

Figure 1-5.
Storm tide and waves from Hurricane Dennis on July 10, 2005, near Panacea, Florida.

SOURCE: U.S. GEOLOGICAL SURVEY (USGS) *SOUND WAVES MONTHLY* NEWSLETTER. PHOTOGRAPH COURTESY OF *THE FORGOTTEN COASTLINE* (COPYRIGHT 2005)

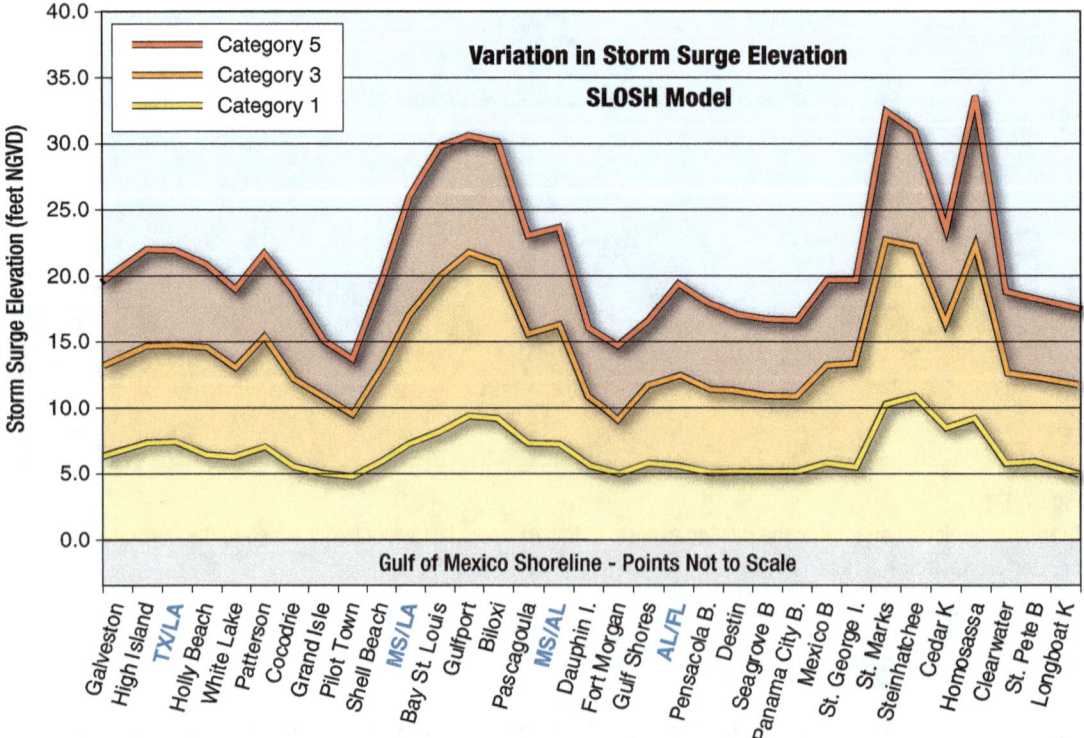

Figure 1-6.
Comparison of storm surge levels along the shorelines of the Gulf Coast for Category 1, 3, and 5 storms.

SOURCE: *HURRICANE KATRINA IN THE GULF COAST* (FEMA 549)

1.3.1 Hydrostatic Forces

Horizontal hydrostatic forces against a structure are created when the level of standing or slowly moving floodwaters on opposite sides of the structure are not equal. Flooding can also cause vertical hydrostatic forces, resulting in flotation. Rapidly rising floodwaters can also cause structures to float off of their foundations (Figure 1-7). If floodwaters rise slowly enough, water can seep into a structure to reduce buoyancy forces. While slowly rising floodwaters reduce the adverse effects of buoyancy, any flooding that inundates a home can cause extreme damage.

Figure 1-7.
Building floated off of its foundation (Plaquemines Parish, Louisiana).

SOURCE: *HURRICANE KATRINA IN THE GULF COAST* (FEMA 549)

1.3.2 Hydrodynamic Forces

Moving floodwaters create hydrodynamic forces on submerged foundations and buildings. These hydrodynamic forces can destroy solid walls and dislodge buildings with inadequate connections or load paths. Moving floodwaters can also move large quantities of sediment and debris that can cause additional damage. In coastal areas, moving floodwaters are usually associated with one or more of the following:

- Storm surge and wave runup flowing landward through breaks in sand dunes, levees, or across low-lying areas (Figure 1-8)

- Outflow (flow in the seaward direction) of floodwaters driven into bay or upland areas by a storm

- Strong currents along the shoreline driven by storm waves moving in an angular direction to the shore

High-velocity flows can be created or exacerbated by the presence of manmade or natural obstructions along the shoreline and by "weak points" formed by shore-normal (i.e., perpendicular to the shoreline) roads and access paths that cross dunes, bridges, or shore-normal canals, channels, or drainage features. For example, evidence after Hurricane Opal (1995) struck Navarre Beach, Florida, suggests that flow was channeled in between large, engineered buildings. The

1 TYPES OF HAZARDS

resulting constricted flow accelerated the storm surge and caused deep scour channels across the island. These channels eventually undermined pile-supported houses between large buildings while also washing out roads and houses farther landward (Figure 1-9).

Figure 1-8.
Aerial view of damage to one of the levees caused by Hurricane Katrina (photo taken on August 30, 2005, the day after the storm hit, New Orleans, Louisiana).

SOURCE: FEMA NEWS PHOTO/ JOCELYN AUGUSTINO

Figure 1-9.
During Hurricane Opal (1995), this house was in an area of channeled flow between large buildings. As a result, the house was undermined and washed into the bay behind a barrier island.

SOURCE: *COASTAL CONSTRUCTION MANUAL* (FEMA 55)

1.3.3 Waves

Waves can affect coastal buildings in a number of ways. The most severe damage is caused by breaking waves (Figures 1-10 and 1-11). The height of these waves can vary by flood zone: V zone wave heights can exceed 3 feet, while Coastal A zone wave heights are between 1.5 and 3 feet. The force created by waves breaking against a vertical surface is often ten or more times higher than the force created by high winds during a storm event. Waves are particularly damaging due to their cyclic nature and resulting repetitive loading. Because typical wave periods during hurricanes range from about 6 to 12 seconds, a structure can be exposed to 300 to 600 waves per hour, resulting in possibly several thousand load cycles over the duration of the storm.

Wave runup occurs as waves break and run up beaches, sloping surfaces, and vertical surfaces. Wave runup can drive large volumes of water against or around coastal buildings, creating hydrodynamic forces (although smaller than breaking wave forces), drag forces from the current, and localized erosion and scour. Wave runup under a vertical surface (such as a wall) will create an upward force by the wave action due to the sudden termination of its flow. This upward force is much greater than the force generated as a wave moves along a sloping surface. In some instances, the force is large enough to destroy overhanging elements such as carports, decks, porches, or awnings. Another negative effect of waves is reflection or deflection occurring when a wave is suddenly redirected as it impacts a building or structure.

Figure 1-10.
Storm waves breaking against a seawall in front of a coastal residence at Stinson Beach, California.

SOURCE: *COASTAL CONSTRUCTION MANUAL* (FEMA 55)

Figure 1-11.
Storm surge and waves overtopping a coastal barrier island in Alabama (Hurricane Frederic, 1979).

SOURCE: *COASTAL CONSTRUCTION MANUAL* (FEMA 55)

1 TYPES OF HAZARDS

1.3.4 Floodborne Debris

Floodborne debris produced by coastal flood events and storms typically includes carports, decks, porches, awnings, steps, ramps, breakaway wall panels, portions of or entire houses, fuel tanks, vehicles, boats, piles, fences, destroyed erosion control structures, and a variety of smaller objects (Figure 1-12). In some cases, larger pieces of floodborne debris can strike buildings (e.g., shipping containers and barges), but the designs contained herein are not intended to withstand the loads from these larger debris elements. Floodborne debris is capable of destroying unreinforced masonry walls, light wood frame construction, and small-diameter posts and piles (and the components of structures they support). Debris trapped by cross-bracing, closely spaced piles, grade beams, or other components is also capable of transferring flood and wave loads to the foundation of an elevated structure.

Figure 1-12.
Pier piles were carried over 2 miles by the storm surge and waves of Hurricane Opal (1995) before coming to rest near this elevated house in Pensacola Beach, Florida.

SOURCE: *COASTAL CONSTRUCTION MANUAL* (FEMA 55)

1.3.5 Erosion and Scour

Erosion refers to the wearing and washing away of coastal lands, including sand and soil. It is part of the larger process of shoreline changes. Erosion occurs when more sediment leaves a shoreline area than enters from either manmade objects or natural forces. Because of the dynamic nature of erosion, it is one of the most complex hazards to understand and difficult to accurately predict at any given site along coastal areas.

Short-term erosion changes can occur from storms and periods of high wave activity, lasting over periods ranging from a few days to a few years. Because of the variability in direction and magnitude, short-term erosion (storm-induced) effects can be orders of magnitude greater than long-term erosion. Long-term shoreline changes occur over a period of decades or longer and tend to average out the short-term erosion. Both short-term and long-term changes should be considered in the siting and design of coastal residential construction. Refer to Chapter 7 of FEMA 55, *Coastal Construction Manual* for additional guidance on assessing short- and long-term erosion.

Scour can occur when water flows at high velocities past an object embedded in or sitting on soil that can be eroded. Scour occurs around the object itself, such as a pile or foundation element, and contributes to the loss of support provided by the soil. In addition to any storm or flood-induced erosion that occurs in the general area, scour is generally limited to small, cone-shaped depressions. Localized scour is capable of undermining slabs, piles, and grade beam structures, and, in severe cases, can lead to structural failure (Figure 1-13). This document considers these effects on the foundation size and depth of embedment requirements.

Figure 1-13.
Extreme case of localized scour undermining a slab-on-grade house in Topsail Island, North Carolina, after Hurricane Fran (1996). Prior to the storm, the lot was several hundred feet from the shoreline and mapped as an A zone on the FIRM. This case illustrates the need for open foundations in Coastal A zones.
SOURCE: *COASTAL CONSTRUCTION MANUAL* (FEMA 55)

FEMA 55 contains guidance on predicting scour, much of which is based on conditions observed after numerous coastal storms. FEMA 55 suggests that scour depths around individual piles be estimated at two times the pile diameter for circular piles and two times the diagonal dimension for square or rectangular piles.

In some storms (e.g., Hurricane Ike, which struck the Texas coast in October 2008), observed scour depths exceed those suggested in FEMA 55. Because erosion and scour reduce the resistance of the pile (by reducing its embedment) and increases stresses within the pile (by increasing the bending moment within the pile that the lateral forces create), they can readily cause failure of a coastal foundation. To help prevent failure, erosion and scour depths should be approximated conservatively. After Hurricane Ike, FEMA developed eight Hurricane Ike Recovery Advisories (RAs). One of the RAs, *Erosion, Scour, and Foundation Design*, discusses erosion and scour and their effects on coastal homes with pile foundations and is presented in Appendix F. All of the Hurricane Ike RAs are available at http://www.fema.gov/library/viewRecord.do?id=3539.

RECOMMENDED RESIDENTIAL CONSTRUCTION
FOR COASTAL AREAS

Building on Strong and Safe Foundations

2. Foundations

This chapter discusses the primary issues related to designing foundations for residential buildings in coastal areas: foundation design criteria, National Flood Insurance Program (NFIP) requirements on coastal construction in A and V zones, the performance of various foundation types, and foundation construction.

2.1 Foundation Design Criteria

Foundations in coastal areas should be designed in accordance with the 2006 or 2009 edition of the IBC or IRC; both contain up-to-date wind provisions and are consistent with NFIP flood provisions. In addition, any locally adopted building ordinances must be addressed. Foundations should be designed and constructed to:

- Properly support the elevated home and resist all loads expected to be imposed on the home and its foundation during a design event

- Prevent flotation, collapse, or lateral movement of the building

2 FOUNDATIONS

- Function after being exposed to the anticipated levels of erosion and scour that may occur over the life of the building.

In addition, the foundation should be constructed with flood-resistant materials below the Base Flood Elevation (BFE).

2.2 Foundation Design in Coastal Areas

Building in a coastal environment is different from building in an inland area because:

- Storm surge, wave action, and erosion in coastal areas make coastal flooding more damaging than inland flooding.
- Design wind speeds are higher in coastal areas and thus require buildings and their foundations to be able to resist higher wind loads.

Foundations in coastal areas must be constructed such that the top of the lowest floor (in A zones) or the bottom of the lowest horizontal structural members (in V zones) of the buildings are elevated above the BFE, while withstanding flood forces, high winds, erosion and scour, and floodborne debris. Deeply embedded pile or other open foundations are required for V zones because they allow waves and floodwaters to pass beneath elevated buildings. Because of the increased flood, wave, floodborne debris, and erosion hazards in V zones, NFIP design and construction requirements are more stringent in V zones than in A zones.

Some coastal areas mapped as A zones may also be subject to damaging waves and erosion (referred to as "Coastal A zones"). A Coastal A zone is also known as the Limit of Moderate Wave Action (LiMWA), which is the landward extent of coastal areas designated Zone AE where waves higher than 1.5 feet can exist during a design flood. Buildings in these areas that are constructed to minimum NFIP A zone requirements may sustain major damage or be destroyed during the base flood. *It is strongly recommended that buildings in A zones subject to breaking waves and erosion be designed and constructed with V zone type foundations* (Figure 2-1). Open foundations are often recommended instead of solid wall, crawlspace, slab, or shallow foundations, which can restrict floodwaters and be undermined easily. Figure 2-2 shows examples of building failures due to erosion and scour under a slab-on-grade foundation.

> **NFIP Minimum Elevation Requirements for New Construction***
>
> A zone: Elevate top of lowest floor to or above BFE
>
> V zone: Elevate bottom of lowest horizontal structural member supporting the lowest floor to or above BFE
>
> In both V and A zones, many property owners have decided to elevate one full story above grade, even if not required, to allow below-building parking. Fact Sheet No. 2 of FEMA 499 contains information about NFIP requirements and recommended best practices in A and V zones (see Appendix F).

---

* For floodplain management purposes, "new construction" means structures for which the start of construction began on or after the effective date of the floodplain management regulation adopted by a community. Substantial improvements, repairs of substantial damage, and some enclosures must meet most of the same requirements as new construction.

FOUNDATIONS 2

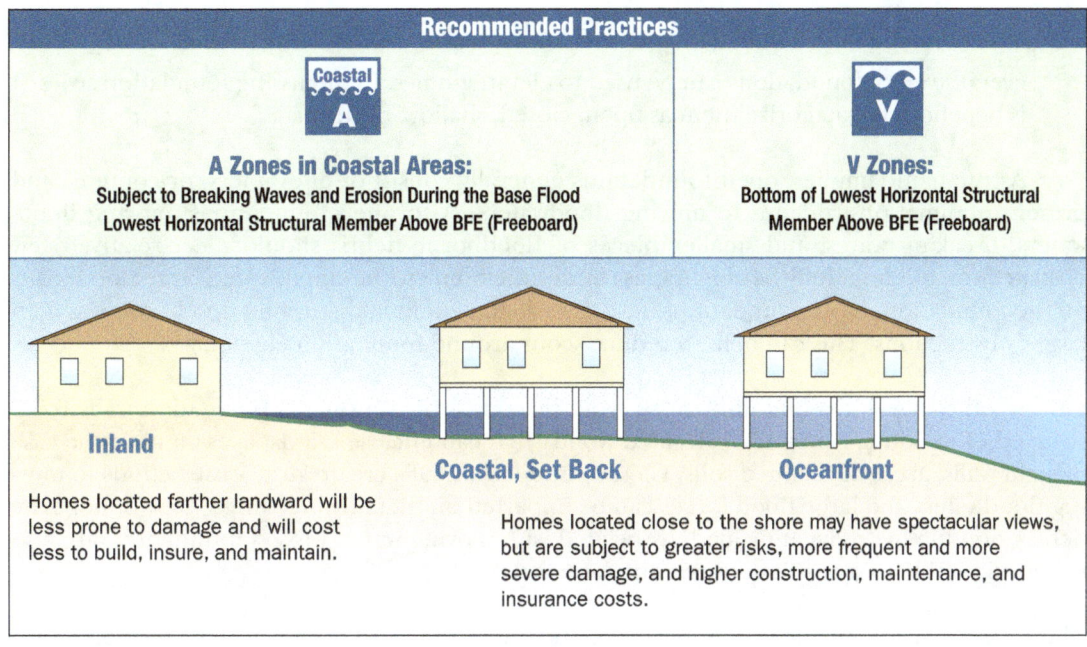

Figure 2-1.
Recommended open foundation practice for buildings in A zones, Coastal A zones, and V zones.

SOURCE: *COASTAL CONSTRUCTION MANUAL* (FEMA 55)

Figure 2-2.
Slab-on-grade foundation failure due to erosion and scour undermining and closeup of the foundation failure from Hurricane Dennis, 2005 (Navarre Beach, Florida).

SOURCE: HURRICANE DENNIS MAT PHOTO

2 FOUNDATIONS

2.3 Foundation Styles in Coastal Areas

Several styles of foundations can be used to elevate homes. In discussing foundation styles, it is beneficial to categorize them as open, closed, shallow, or deep.

As the name implies, open foundations generally consist of piles, piers, or columns and present minimal obstructions to moving floodwaters. With open foundations, moving floodwaters, breaking waves, and smaller pieces of floodborne debris should meet relatively few obstructions and hopefully be able to pass under the home without imparting large flood loads on the foundation. Open foundations have the added benefit of disrupting flood flows less than larger obstructions. This can help to reduce scour around foundation elements.

On the other hand, closed foundations typically consist of continuous foundation walls (constructed of masonry, concrete, or treated wood) that can enclose crawlspaces or, as in the case of stem walls, areas of retained soils. Closed foundation walls create large obstructions to moving floodwaters and large flood forces can be imparted on them by breaking waves, floodborne debris, and the hydrodynamic loads associated with moving water. Closed foundations are also more vulnerable to scour than open foundations.

The terms shallow and deep signify the relative depth of the soils on which the homes are founded. Shallow foundations are set on soils that are relatively close to the surface of the surrounding grade, generally within 3 feet of the finished grade. In cold climates, shallow foundations may need to be extended 4 feet or more below grade to set the foundation beneath the design frost depth. Shallow foundations can consist of discrete concrete pad footings, strip footings, or a matrix of strip footings placed to create a mat foundation. Mat foundations have the added benefit of better resisting uplift and overturning forces than foundations consisting of discrete pad footings.

Deep foundations are designed to be supported on much deeper soils or rock. These foundations frequently are used where soils near the surface have relatively weak bearing capacities (typically 700 pounds per square foot [psf] or less), when soils near the surface contain expansive clays (also called shrink/swell soils because they shrink when dry and swell when wet) or where surface soils are vulnerable to being removed by erosion or scour.

Although typical foundation styles vary geographically, deep foundations for residential construction in coastal areas generally consist of driven treated timber piles or treated square piles. Driven concrete piles are common in other areas.

Only open foundations with base members or elements (piles or beams) located below expected erosion and scour are allowed in V zones; as a "best practices" approach, open foundations are recommended, but are not NFIP required, in Coastal A zones. Table 2-1 shows the recommended type of foundation depending on the coastal area. Additional information concerning foundation performance in coastal areas can be found in FEMA 499, Fact Sheet No. 11 (see Appendix F).

FOUNDATIONS 2

Table 2-1. Foundation Type Dependent on Coastal Area

Foundation Type	V Zone	Coastal A Zone	A Zone
Open	✔	✔	✔
Closed	✘	NR	✔

✔ = Acceptable   NR = Not Recommended   ✘ = Not Permitted

2.3.1 Open Foundations

Open foundations are required in V zones and recommended in Coastal A zones. As previously mentioned, this type of foundation allows water to pass beneath an elevated building through the foundation and reduces lateral flood loads on the structure. Open foundations also have the added benefit of being less susceptible to damage from floodborne debris because debris is less likely to be trapped.

2.3.1.1 Piles

Pile foundations consist of deeply placed vertical piles installed under the elevated structure. The piles support the elevated structure by remaining solidly placed in the soil. Because pile foundations are set deeply, they are inherently more tolerant to erosion and scour. Piles rely primarily on the friction forces that develop between the pile and the surrounding soils (to resist gravity and uplift forces) and the compressive strength of the soils (to resist lateral movement). The soils at the ends of the piles also contribute to resist gravity loads.

Piles are typically treated wood timbers, steel pipes, or pre-cast concrete. Other materials like fiber reinforced polyester (FRP) are available, but are rarely used in residential construction. Piles can be used with or without grade beams. When used without grade beams, piles extend to the lowest floor of the elevated structure. Improved performance is achieved when the piles extend beyond the lowest floor to the roof (or an upper floor level) above. Doing so provides resistance to rotation (also called "fixity") in the top of the pile and improves stiffness of the pile foundation. Occasionally, wood framing members are installed at the base of a wood pile (Figure 2-3). These members are not true grade beams but rather are compression struts. They provide lateral support for portions of the pile near grade and reduce the potential for column buckling; however, due to the difficulties of constructing moment connections with wood, the compression struts provide very little resistance to rotation.

Critical aspects of a pile foundation include the pile size, installation method, and embedment depth, bracing, and their connections to the elevated structure (see FEMA 499, Fact Sheet Nos. 12 and 13 in Appendix F). Pile foundations with inadequate embedment will not have the structural capacity to resist sliding and overturning (Figure 2-4). Inadequate embedment and improperly sized piles greatly increase the probability for structural collapse. However, when properly sized, installed, and braced with adequate embedment into the soil (with consideration for erosion and scour effects), a building's pile foundation performance will allow the building to remain standing and intact following a design flood event (Figure 2-5).

2 FOUNDATIONS

Figure 2-3.
Compression strut at base of a wood pile. Struts provide some lateral support for the pile, but very little resistance to rotation.

SOURCE: *COASTAL CONSTRUCTION MANUAL* (FEMA 55)

Figure 2-4.
Near collapse due to insufficient pile embedment (Dauphin Island, Alabama).

SOURCE: *HURRICANE KATRINA IN THE GULF COAST* (FEMA 549)

Figure 2-5.
Successful pile foundation following Hurricane Katrina (Dauphin Island, Alabama).

SOURCE: *HURRICANE KATRINA IN THE GULF COAST* (FEMA 549)

When used with grade beams, the piles and grade beams work in conjunction to elevate the structure, provide vertical and lateral support for the elevated home, and transfer loads imposed on the elevated home and foundation to the ground below.

Pile foundations with grade beams must be constructed with adequate strength to resist all lateral and vertical loads. Failures experienced during Hurricane Katrina often resulted from inadequate connections between the columns and footings or grade beams below (Figure 2-6). Pile and grade beam foundations should be designed and constructed so that the grade beams act only to provide fixity to the foundation system and not to support the lowest elevated floor. If grade beams support the lowest elevated floor of the home, they become the lowest horizontal structural member and significantly higher flood insurance premiums would result. Also, if the grade beams support the structure, the structure would become vulnerable to erosion and scour. Grade beams must also be designed to span between adjacent piles and the piles must be capable of resisting both the weight of the grade beams when undermined by erosion and scour, and the loads imposed on them by forces acting on the structure.

Figure 2-6.
Column connection failure (Belle Fontaine Point, Jackson County, Mississippi).

SOURCE: *HURRICANE KATRINA IN THE GULF COAST* (FEMA 549)

2.3.1.2 Piers

Piers are generally placed on footings to support the elevated structure. Without footings, piers function as short piles and rarely have sufficient capacity to resist uplift and gravity loads. The type of footing used in pier foundations greatly affects the foundation's performance (Figure 2-7). When exposed to lateral loads, discrete footings can rotate so piers placed on discrete footings are only suitable when wind and flood loads are relatively low. Piers placed on continuous concrete grade beams or concrete strip footings provide much greater resistance to lateral loads because the grade beams/footings act as an integral unit and are less prone to rotation. Footings and grade beams must be reinforced to resist the moment forces that develop at the base of the piers due to the lateral loads on the foundation and the elevated home (Figure 2-8).

2 FOUNDATIONS

Since pier foundation footings or grade beams are limited in depth of placement, they are appropriate only where there is limited potential for erosion or scour. The maximum estimated depth for long- and short-term erosion and localized scour should not extend below the bottom of the footing or grade beam.

Figure 2-7. Performance comparison of pier foundations. Piers on discrete footings (foreground) failed by rotating and overturning while piers on more substantial footings (in this case a concrete mat) survived (Pass Christian, Mississippi).

SOURCE: HURRICANE KATRINA MAT PHOTO

2.3.2 Closed Foundations

A closed foundation is typically constructed using foundation walls, a crawlspace foundation, or a stem wall foundation (usually filled with compacted soil). A closed foundation does not allow water to pass easily through the foundation elements below the elevated building. Thus, these types of foundations are said to obstruct the flow. These foundations also present a large surface area upon which waves and flood forces act; therefore, they are prohibited in V zones and not recommended for Coastal A zones. If foundation or crawlspace walls enclose space below the Base Flood Elevation (BFE), they must be equipped with openings that allow floodwaters to flow in and out of the area enclosed by the walls (Figure 2-9 presents an isometric view). The entry and exit of floodwaters will equalize the water pressure on both sides of the wall and reduce the likelihood of the wall collapsing (see FEMA 499, Fact Sheet No. 15 in Appendix F). Two types of closed foundations are discussed in this manual, perimeter walls and slab-on-grade.

2.3.2.1 Perimeter Walls

Perimeter walls are conventional walls (typically masonry or wood frame) that extend from the ground up to the elevated building. They typically bear on shallow footings. Crawlspaces and stem walls are two types of foundations with perimeter walls.

FOUNDATIONS 2

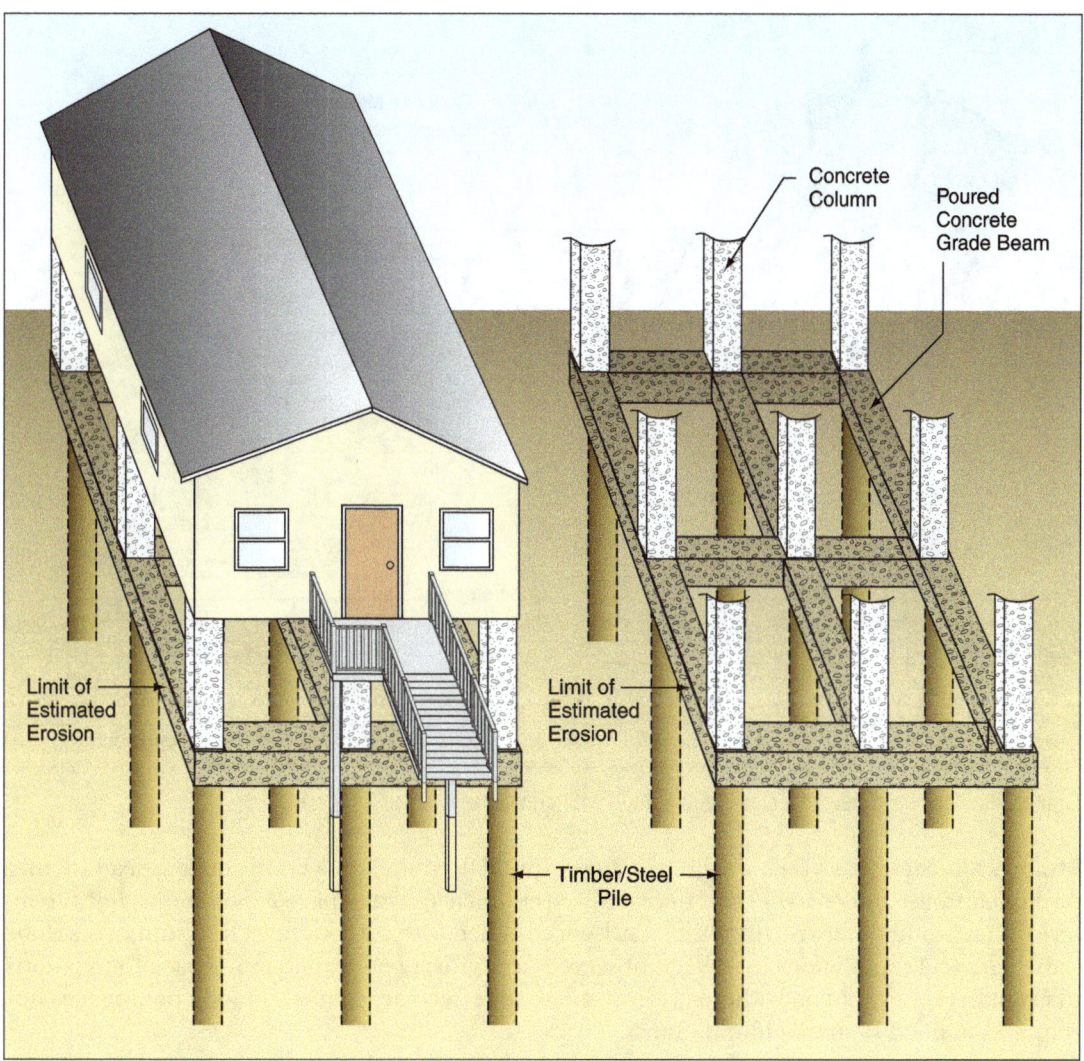

Figure 2-8. Isometric view of an open foundation with grade beam.

Crawlspaces. Crawlspace foundations are typically low masonry perimeter walls, some requiring interior piers supporting a floor system if the structure is wide. These foundations are usually supported by shallow footings and are prone to failure caused by erosion and scour.

This type of foundation is characterized by a solid perimeter foundation wall around a structure with a continuous spread footing with reinforced masonry or concrete piers. All crawlspace foundation walls in the Special Flood Hazard Area (SFHA) must be equipped with flood openings. These openings are required to equalize the pressure on either side of the wall (see FEMA 499, Fact Sheet Nos. 15 and 26 in Appendix F). However, even with flood vents, hydrodynamic and wave forces in Coastal A zones can damage or destroy these foundations.

2 FOUNDATIONS

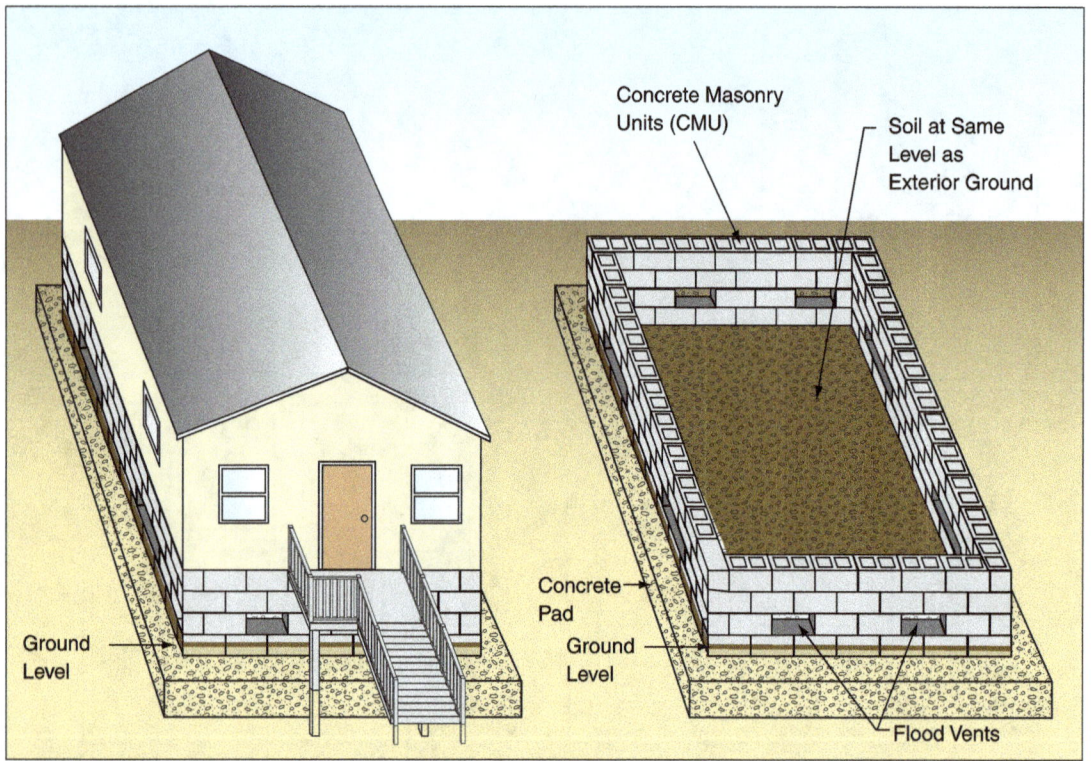

Figure 2-9. Isometric view of a closed foundation with crawlspace.

Stem Walls. Stem walls (i.e., a solid perimeter foundation wall on a continuous spread footing backfilled to the underside of the floor slab) are similar to crawlspace foundations, but the interior that would otherwise form the crawlspace is filled with soil or gravel that supports a floor slab. Stem wall foundations have been observed to perform better than crawlspace foundations in Coastal A zones (but only where erosion and scour effects are minor). Flood openings are not required in filled stem wall foundations.

2.3.2.2 Slab-on-Grade

A slab-on-grade foundation is concrete placed directly on-grade (to form the slab) with generally thickened, reinforced sections around the edges and under loadbearing walls. The slab itself is typically 4 inches thick where not exposed to concentrated loads and 8 to 12 inches thick under loadbearing walls. The thickened portions of slab-on-grade foundations are typically reinforced with deformed steel bars to provide structural support; the areas not thickened are typically reinforced with welded wire fabrics (WWFs) for shrinkage control. While commonly used in residential structures in A zones, slab-on-grade foundations are prone to erosion, prohibited in V zones, and not recommended for Coastal A zones.

Slab-on-grade foundations can be used with structural fill to elevate buildings. Fill is usually placed in layers called "lifts" with each lift compacted at the site. Because fill is susceptible to erosion, it is prohibited for providing structural support in V zones. Structural fill is not recommended for Coastal A zones, but may be appropriate for non-Coastal A zones.

2.4 Introduction to Foundation Design and Construction

This section introduces two main issues related to foundation design and construction: site characterization and types of foundation construction. Construction materials and methods are addressed in Chapter 4.

2.4.1 Site Characterization

The foundation design chosen should be based on the characteristics that exist at the building site. A site characteristic study should include the following:

- The type of foundations that have been installed in the area in the past. A review of the latest FIRM is recommended to ensure that construction characteristics have not been changed.

- The proposed site history, which would indicate whether there are any buried materials or if the site has been regraded.

- How the site may have been used in the past, from a search of land records for past ownership.

- A soil investigation report, which should include:
    - Soil borings sampled from the site or taken from test pits
    - A review of soil borings from the immediate area adjacent to the site
    - Information from the local office of the Natural Resource Conservation Service (NRCS) (formerly the Soil Conservation Service [SCS]) and soil surveys published for each county

One of the parameters derived from a soil investigation report is the bearing capacity, which measures the ability of soils to support gravity loads without soil shear failure or excessive settlement. Measured in psf, soil bearing capacity typically ranges from 1,000 psf for relatively weak soils to over 10,000 psf for bedrock.

Frequently, designs are initially prepared on a presumed bearing capacity. It is then the homebuilder's responsibility to verify actual site conditions. The actual soil bearing capacity should be determined. If soils are found to have higher bearing capacity, the foundation can be constructed as designed or the foundation can be revised to take advantage of the better soils.

Allowable load bearing values of soils given in Section 1806 of the 2009 IBC can be used when other data are not available. However, soils can vary significantly in bearing capacity from one site to the next. A geotechnical engineer should be consulted when any unusual or unknown soil condition is encountered.

2 FOUNDATIONS

2.4.2 Types of Foundation Construction

2.4.2.1 Piles

A common type of pile foundation is the elevated wood pile foundation, where the piles extend from deep in the ground to an elevation at or above the Design Flood Elevation (DFE). Horizontal framing members are used to connect the piles in both directions. This grid forms a platform on which the house is built (see FEMA 499, Fact Sheet No. 12 in Appendix F).

The method of installation is a major consideration in the structural integrity of pile foundations. The ideal method is to use a pile driver. In this method, the pile is held in place with leads while a single-acting, double-acting diesel, or air-powered hammer drives the pile into the ground (Figure 2-10).

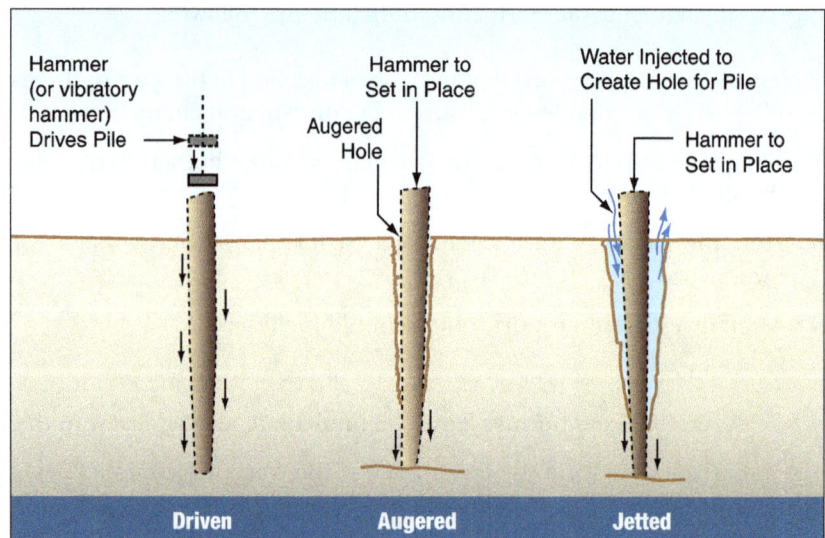

Figure 2-10. Pile installation methods.

SOURCE: *COASTAL CONSTRUCTION MANUAL* (FEMA 55)

If steel piles are used, only the hammer driving method mentioned above should be used. For any pile driving, the authority having jurisdiction or the engineer-of-record may require that a driving log is kept for each pile. The log will tabulate the number of blows per foot as the driving progresses. This log is a key factor used in determining the pile capacity.

Another method for driving piles is the drop hammer method. It is a lower cost alternative to the pile driver. A drop hammer consists of a heavy weight raised by a cable attached to a power-driven winch and then dropped onto the pile.

Holes for piles may be excavated by an auger if the soil has sufficient clay or silt content. Augering can be used by itself or in conjunction with pile driving. If the hole is full-sized, the pile is dropped in and the void backfilled. Alternatively, an undersized hole can be drilled and a pile driven into it. When the soil conditions are appropriate, the hole will stay open long enough to drop or drive in a pile. In general, this method may not have as much capacity as those methods previously mentioned. Like jetted piles, augered piles are not appropriate for the designs provided in this manual unless the method for compressing the soil is approved by a geotechnical engineer.

A less desirable but frequently used method of inserting piles into sandy soil is "jetting," which involves forcing a high-pressure stream of water through a pipe that advances with the pile. The water creates a hole in the sand as the pile is driven until the required depth is reached. Unfortunately, jetting loosens the soil both around the pile and the tip. This results in a lower load capacity due to less frictional resistance. Jetted piles are not appropriate for the designs provided in this manual unless capacity is verified by a geotechnical engineer.

2.4.2.2 Diagonal Bracing of Piles

The foundation design may include diagonal bracing to stiffen the pile foundation in one or more directions. When installed properly, bracing lowers the point where lateral loads are applied to the piles. The lowering of load application points reduces the bending forces that piles must resist (so piles in a braced pile foundation do not need to be as strong as piles in an unbraced pile foundation) and also reduces lateral movement in the building. Outside piles are sufficiently designed to withstand external forces, because bracing will not assist in countering these forces. A drawback to bracing, however, is that the braces themselves can become obstructions to moving floodwaters and increase a foundation's exposure to wave and debris impact.

Because braces tend to be slender, they are vulnerable to compression buckling. Therefore, most bracing is considered tension-only bracing. Because wind and, to a lesser extent, flood loads can act in opposite directions, tension-only bracing must be installed in pairs. One set of braces resists loads from one direction while the second set resists loads from the opposite direction. Figure 2-11 shows how tension only bracing pairs resist lateral loads on a home.

The braced pile design can only function when all of the following conditions are met:

- The home must be constructed with a stiff horizontal diaphragm such as a floor system that transfers loads to laterally braced piles.
- Solid connections, usually achieved with bolts, must be provided to transmit forces from the brace to the pile or floor system.

The placement of the lower bolted connection of the diagonal to the pile requires some judgment. If the connection is placed too high above grade, the pile length below the connection is not braced and the overall bracing is less strong and stiff. If the connection is placed too close to grade, the bolt hole is more likely to be flooded or infested with termites. Because the bolt hole passes through the untreated part of the pile, flooding and subsequent decay or termite infestation will weaken the pile at a vulnerable location. Therefore, the bolt hole should be treated with preservative after drilling and prior to bolt placement.

The braced wood pile designs developed for this manual use steel rods for bracing. Steel rods were used because:

- Steel has greater tensile strength than even wide dimensional lumber.

2 FOUNDATIONS

- There are fewer obstructions to waves and floodborne debris.
- The rod bracing can easily be tensioned with turnbuckles and can be adjusted throughout the life of the home.
- A balanced double shear connection is two to three times stronger than a wood to wood connection made with 2-inch thick dimensional lumber.

Alternative bracing should only be installed when designed by a licensed engineer.

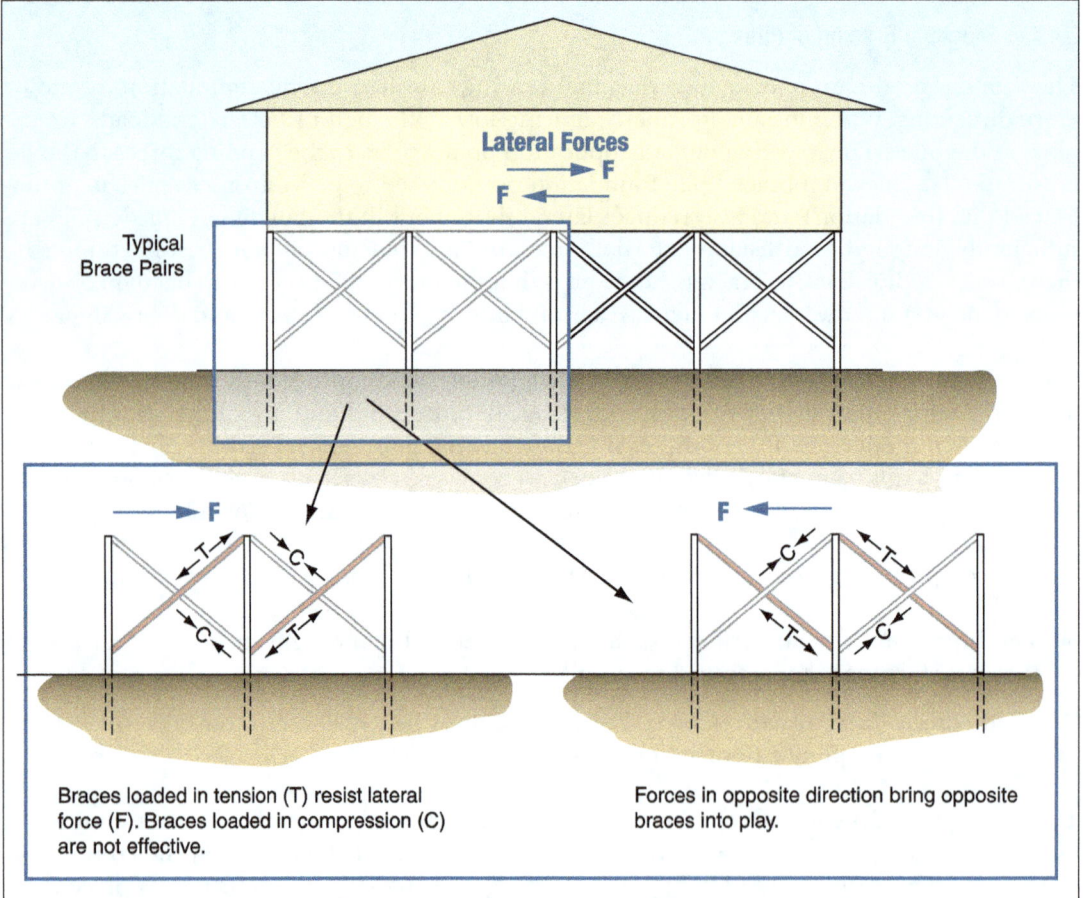

Figure 2-11. Diagonal bracing schematic.

2.4.2.3 Knee Bracing of Piles

Knee braces involve installing short diagonal braces between the upper portions of the piles and the floor system of the elevated structure. The braces increase the stiffness of an elevated pile foundation and can be effective at resisting the lateral forces on a home. Although knee braces do not stiffen a foundation as much as diagonal bracing, they do offer some advantages over diagonal braces. For example, knee braces present less obstruction to waves and debris, are shorter than diagonal braces, and are usually designed for both tension and compression loads. Unlike diagonal braces, knee braces do not reduce bending moments in the piles (they can actually increase bending moments) and will not reduce the diameter of the piles required to resist lateral loads.

The entire load path into and through the knee brace must be designed with sufficient capacity. The connections at each end of each knee brace must have sufficient capacity to handle both tension and compression and to resist axial loads in the brace. The brace itself must have sufficient cross-sectional area to resist compression and tensile loads. Due to the complexity of knee bracing, they have not been used in the foundation designs included in Appendix A herein.

2.4.2.4 Wood-Pile-to-Wood-Girder Connections

Wood piles are often notched to provide a bearing surface for a girder. However, a notch should not reduce more than 50 percent of the pile cross-section (such information is typically provided by a designer on contract documents). For proper load transfer, the girder should bear on the surface of the pile notch.

Although connections play an integral role in the design of structures, they are typically regarded as the weakest link. The connection between a wood pile and the elevated structure should be designed by a licensed engineer (see FEMA 499, Fact Sheet No. 13 in Appendix F).

2.4.2.5 Grade Beams in Pile/Column Foundations

Grade beams are sometimes used in conjunction with pile and column foundations to generate more stiffness. They generate stiffness by forcing the piles to move as a group rather than individually and by providing fixity (i.e., resistance to rotation) at the ends of the piles. Typically, they extend in both directions and are usually made of reinforced concrete. The mix design, the amount and placement of reinforcement, the cover, and the curing process are important parameters in optimizing durability. To reduce the effect of erosion and scour on foundations, grade beams must be designed to be self-supporting foundation elements. The supporting piers should be designed to carry the weight of the grade beams and resist all loads transferred to the piers.

In V zones, grade beams must be used only for lateral support of the piles. If, during construction, the floor is made monolithic with the grade beams, the bottom of the beams become the lowest horizontal structural member. This elevation must be at or above the BFE.

If grade beams are used with wood piles, the possibility of rot occurring must be considered when designing the connection between the grade beam and the pile. The connection must not encourage water retention. The maximum bending moment in the piles occurs at the grade beams, and decay caused by water retention at critical points in the piles could induce failure under high-wind or flood forces.

RECOMMENDED RESIDENTIAL CONSTRUCTION
FOR COASTAL AREAS

Building on Strong and Safe Foundations

3. Foundation Design Loads

This chapter provides guidance on how to determine the magnitude of the loads placed on a building by a particular natural hazard event or a combination of events. The methods presented are intended to serve as the basis of a methodology for applying the calculated loads to the building during the design process.

The process for determining site-specific loads from natural hazards begins with identifying the building codes or engineering standards in place for the selected site (e.g., the International Building Code 2009 (IBC 2009) or ASCE 7-05, *Minimum Design Loads for Buildings and Other Structures*), if model building codes and other building standards do not provide load determination and design guidance for each of the hazards identified. In these instances, supplemental guidance such as FEMA 55 should be sought, the loads imposed by each of the identified hazards should be calculated, and the load combinations appropriate for the building site should be determined. The load combinations used in this manual are those specified by ASCE 7-05, the standard referenced by the IBC 2009. Either allowable stress design (ASD) or strength

3 FOUNDATION DESIGN LOADS

design methods can be used to design a building. For this manual, all of the calculations, analyses, and load combinations presented are based on ASD. The use of strength design methods will require the designer to modify the design values to accommodate strength design concepts. Assumptions utilized in this manual can be found in Appendix C.

3.1 Wind Loads

Wind loads on a building structure are calculated using the methodology presented in ASCE 7-05. This document is the wind standard referenced by the 2003 editions of the IBC and IRC. Equations used to calculate wind loads are presented in Appendix D.

The most important variable in calculating wind load is the design wind speed. Design wind speed can be obtained from the local building official or the ASCE 7-05 wind speed map (Figure 3-1). The speeds shown in this figure are 3-second gust speeds for Exposure Category C at a 33-foot (10-meter) height. ASCE 7-05 includes scaling factors for other exposures and heights.

ASCE 7-05 specifies wind loads for structural components known as a main wind force resisting system (MWFRS). The foundation designs developed for this manual are based on MWFRS pressures calculated for Exposure Category C, the category with the highest anticipated wind loads for land-based structures.

ASCE 7-05 also specifies wind loads for components and cladding (C&C). Components and cladding are considered part of the building envelope, and ASCE 7-05 requires C&C to be designed to resist higher wind pressures than a MWFRS.

3.2 Flood Loads

This manual develops in more detail flood load calculations and incorporates the methodology presented in ASCE 7-05. Although wind loads can directly affect a structure and dictate the actual foundation design, the foundation is more affected by flood loads. ASCE 24-05 discusses floodproof construction. Loads developed in ASCE 24-05 come directly from ASCE 7-05, which is what the designs presented herein are based upon.

The effects of flood loads on buildings can be exacerbated by storm-induced erosion and localized scour, and by long-term erosion. Erosion and scour lower the ground surface around foundation members and can cause the loss of load-bearing capacity and resistance to lateral and uplift loads. Erosion and scour also increase flood depths and, therefore, increase depth dependent flood loads.

3.2.1 Design Flood and DFE

The design flood is defined by ASCE 7-05 as the greater of the following two flood events:

1. Base flood, affecting those areas identified as SFHAs on the community's FIRM, or

FOUNDATION DESIGN LOADS 3

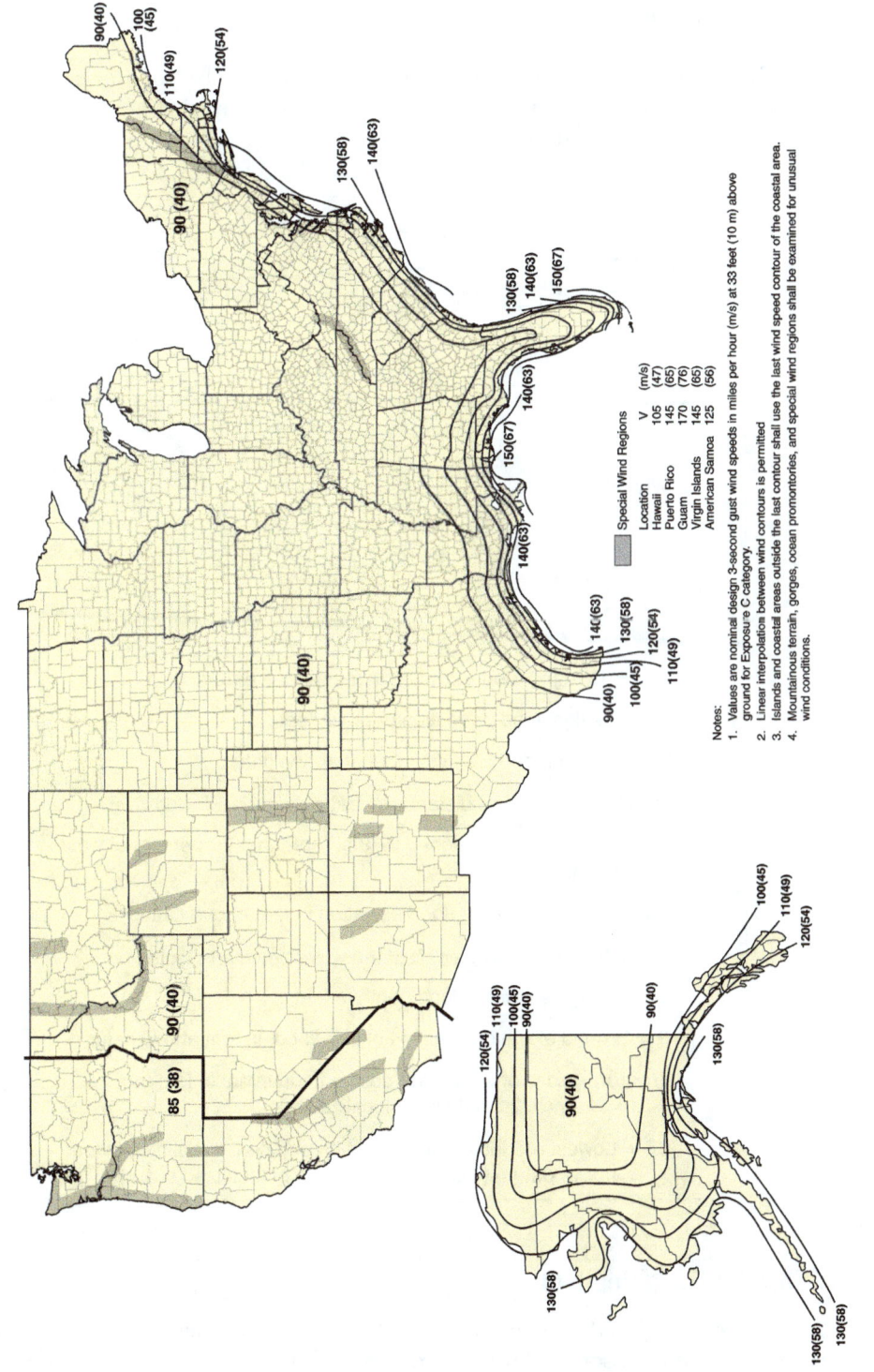

Figure 3-1. Wind speeds (in mph) for the entire U.S.

SOURCE: ASCE 7-05

3 FOUNDATION DESIGN LOADS

2. The flood corresponding to the area designated as a flood hazard area on a community's flood hazard map or otherwise legally designated.

The DFE is defined as the elevation of the design flood, including wave height and freeboard, relative to the datum specified on a community's flood hazard map. Figure 3-2 shows the parameters that determine or are affected by flood depth.

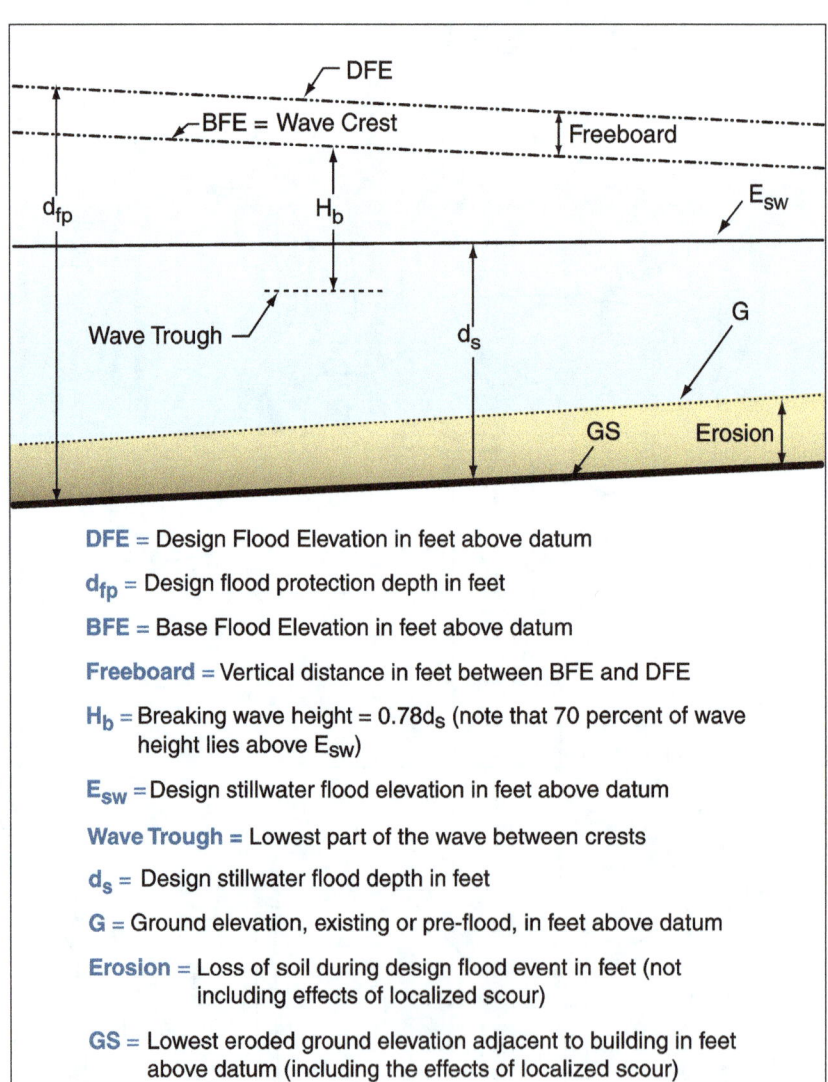

Figure 3-2. Parameters that determine or are affected by flood depth.

SOURCE: *COASTAL CONSTRUCTION MANUAL* (FEMA 55)

DFE = Design Flood Elevation in feet above datum

d_{fp} = Design flood protection depth in feet

BFE = Base Flood Elevation in feet above datum

Freeboard = Vertical distance in feet between BFE and DFE

H_b = Breaking wave height = $0.78 d_s$ (note that 70 percent of wave height lies above E_{sw})

E_{sw} = Design stillwater flood elevation in feet above datum

Wave Trough = Lowest part of the wave between crests

d_s = Design stillwater flood depth in feet

G = Ground elevation, existing or pre-flood, in feet above datum

Erosion = Loss of soil during design flood event in feet (not including effects of localized scour)

GS = Lowest eroded ground elevation adjacent to building in feet above datum (including the effects of localized scour)

3.2.2 Design Stillwater Flood Depth (d_s)

Design stillwater flood depth (d_s) is the vertical distance between the eroded ground elevation and the stillwater flood elevation associated with the design flood. Determining the maximum design stillwater flood depth over the life of a building is the single most important flood load calculation that will be made; nearly all other coastal flood load parameters or calculations (e.g., hydrostatic load, design flood velocity, hydrodynamic load, design wave height, DFE, debris impact load, local scour depth) depend directly or indirectly on the design stillwater flood depth. The design stillwater flood depth (d_s) is defined as:

$$d_s = E_{sw} - GS$$

Where

d_s = Design stillwater flood depth (ft)

E_{sw} = Design stillwater flood elevation (ft) above the datum (e.g., National Geodetic Vertical Datum [NGVD], North American Vertical Datum [NAVD]), including wave setup effects

GS = Lowest eroded ground elevation above datum (ft), adjacent to building, including the effects of localized sour around piles

GS is not the lowest existing pre-flood ground surface; it is the lowest ground surface that will result from long-term erosion and the amount of erosion expected to occur during a design flood, excluding local scour effects. The process for determining GS is described in Chapter 7 of FEMA 55.

Values for E_{sw} are not shown on a FIRM, but they are given in the Flood Insurance Study (FIS) report, which is produced in conjunction with the FIRM for a community. FIS reports are usually available from community officials, from NFIP State Coordinating Agencies, and on the web at the FEMA Map Service Center (http://store.msc.fema.gov). Some States have FIS reports available on their individual web sites.

3.2.3 Design Wave Height (H_b)

The design wave height at a coastal building site will be one of the most important design parameters. Therefore, unless detailed analysis shows that natural or manmade obstructions will protect the site during a design event, wave heights at a site will be calculated from Equation 5-2 of ASCE 7-05 as the heights of depth-limited breaking waves (H_b), which are equivalent to 0.78 times the design stillwater flood depth:

$$H_b = 0.78 d_s$$

Note: 70 percent of the breaking wave height ($0.7 H_b$) lies above the stillwater flood level.

3.2.4 Design Flood Velocity (V)

Estimating design flood velocities in coastal flood hazard areas is subject to considerable uncertainty. Little reliable historical information exists concerning the velocity of floodwaters during coastal flood events. The direction and velocity of floodwaters can vary significantly throughout a coastal flood event, approaching a site from one direction during the beginning of the flood event before shifting to another (or several directions). Floodwaters can inundate some low-lying coastal sites from both the front (e.g., ocean) and the back (e.g., bay, sound, river). In a similar manner, flow velocities can vary from close to zero to high velocities during a single flood event. For these reasons, flood velocities should be estimated conservatively by assuming that floodwaters can approach from the most critical direction and that flow velocities can be high.

For design purposes, the *Commentary* of ASCE 7-05 suggested a range of flood velocities from:

$$V = d_s \div t \text{ (expected lower bound)}$$

to

$$V = (gd_s)^{0.5} \text{ (expected upper bound)}$$

Where

V = Average velocity of water in ft/s

d_s = Design stillwater flood depth

t = Time (1 second)

g = Gravitational constant (32.2 ft/sec^2)

Factors that should be considered before selecting the upper- or lower-bound flood velocity for design include:

- Flood zone
- Topography and slope
- Distance from the source of flooding
- Proximity to other buildings or obstructions

The upper bound should be taken as the design flood velocity if the building site is near the flood source, in a V zone, in an AO zone adjacent to a V zone, in an A zone subject to velocity flow and wave action, steeply sloping, or adjacent to other buildings or obstructions that will confine floodwaters and accelerate flood velocities. The lower bound is a more appropriate design flood velocity if the site is distant from the flood source, in an A zone, flat or gently sloping, or unaffected by other buildings or obstructions.

3.3 Hydrostatic Loads

Hydrostatic loads occur when standing or slowly moving water comes into contact with a building or building component. These loads can act laterally (pressure) or vertically (buoyancy).

FOUNDATION DESIGN LOADS   3

Lateral hydrostatic forces are generally not sufficient to cause deflection or displacement of a building or building component unless there is a substantial difference in water elevation on opposite sides of the building or component; therefore, the NFIP requires that floodwater openings be provided in vertical walls that form an enclosed space below the BFE for a building in an A zone.

Lateral hydrostatic force is calculated by the following:

$$f_{stat} = \tfrac{1}{2}\, \gamma\, d_s^2$$

Where

f_{stat} = Hydrostatic force per unit width (lb/ft) resulting from flooding against vertical element

γ = Specific weight of water (62.4 lb/ft^3 for freshwater and 64 lb/ft^3 for saltwater)

Vertical hydrostatic forces during design flood conditions are not generally a concern for properly constructed and elevated coastal buildings. Buoyant or flotation forces on a building can be of concern if the actual stillwater flood depth exceeds the design stillwater flood depth.

Vertical (buoyancy) hydrostatic force is calculated by the following:

$$F_{Buoy} = \gamma\, (Vol)$$

Where

F_{Buoy} = vertical hydrostatic force (lb) resulting from the displacement of a given volume of floodwater

Vol = volume of floodwater displaced by a submerged object (ft^3) = displaced area x depth of flooding

Buoyant force acting on an object must be resisted by the weight of the object and any other opposing force (e.g., anchorage forces) resisting flotation. In the case of a building, the live load on floors should not be counted on to resist buoyant forces.

3.4 Wave Loads

Calculating wave loads requires information about expected wave heights. For the purposes of this manual, the calculations will be limited by water depths at the site of interest. Wave forces can be separated into four categories:

- Non-breaking waves (can usually be computed as hydrostatic forces against walls and hydrodynamic forces against piles)

3 FOUNDATION DESIGN LOADS

- Breaking waves (short duration but large magnitude forces against walls and piles)
- Broken waves (similar to hydrodynamic forces caused by flowing or surging water)
- Uplift (often caused by wave runup, deflection, or peaking against the underside of horizontal surfaces)

Of these four categories, the forces from breaking waves are the largest and produce the most severe loads. Therefore, it is strongly recommended that the breaking wave load be used as the design wave load.

Two breaking wave loading conditions are of interest in residential construction: waves breaking on small-diameter vertical elements below the DFE (e.g., piles, columns in the foundation of a building in a V zone) and waves breaking against vertical walls below the DFE (e.g., solid foundation walls in A zones, breakaway walls in V zones).

3.4.1 Breaking Wave Loads on Vertical Piles

The breaking wave load (F_{brkp}) on a pile can be assumed to act at the stillwater flood level and is calculated by Equation 5-4 from ASCE 7-05:

$$F_{brkp} = (1/2)C_D \gamma D H_b^2$$

Where

F_{brkp} = Net wave force (lb)

C_D = Coefficient of drag for breaking waves = 1.75 for round piles or column, and 2.25 for square piles or columns

γ = Specific weight of water (lb/ft^3)

D = Pile or column diameter (ft) for circular section. For a square pile or column, 1.4 times the width of the pile or column (ft).

H_b = Breaking wave height (ft)

3.4.2 Breaking Wave Loads on Vertical Walls

The net force resulting from a normally incident breaking wave (depth limited in size, with $H_b = 0.78d_s$) acting on a rigid vertical wall, can be calculated by Equation 5-6 from ASCE 7-05:

$$F_{brkw} = 1.1 C_p \gamma d_s^2 + 2.4 \gamma d_s^2$$

Where

F_{brkw} = net breaking wave force per unit length of structure (lb/ft) acting near the stillwater flood elevation

C_p = Dynamic pressure coefficient ($1.6 < C_p < 3.5$) (see Table 3-1)

Table 3-1. Building Category and Corresponding Dynamic Pressure Coefficient (C_p)

Building Category	C_p
I – Buildings and other structures that represent a low hazard to human life in the event of a failure	1.6
II – Buildings not in Category I, III, or IV	2.8
III – Buildings and other structures that represent a substantial hazard to human life in the event of a failure	3.2
IV – Buildings and other structures designated as essential facilities	3.5

SOURCE: ASCE 7-02

γ = Specific weight of water (lb/ft^3)

d_s = Design stillwater flood depth (ft) at base of building where the wave breaks

This formula assumes the following:

- The vertical wall causes a reflected or standing wave against the seaward side of the wall with the crest of the wave, reaching a height of 1.2d_s above the design stillwater flood elevation, and

- The space behind the vertical wall is dry, with no fluid balancing the static component of the wave force on the outside of the wall (Figure 3-3).

If free-standing water exists behind the wall (Figure 3-4), a portion of the hydrostatic component of the wave pressure and force disappears and the net force can be computed using Equation 5-7 from ASCE 7-05:

$$F_{brkw} = 1.1 C_p \gamma d_s^2 + 1.9 \gamma d_s^2$$

Post-storm damage inspections show that breaking wave loads have virtually destroyed all wood frame or unreinforced masonry walls below the wave crest elevation; only highly engineered, massive structural elements are capable of withstanding breaking wave loads. Damaging wave pressures and loads can be generated by waves much lower than the 3-foot wave currently used by FEMA to distinguish between A and V zones.

3.5 Hydrodynamic Loads

Water flowing around a building (or a structural element or other object) imposes additional loads on the building. The loads (which are a function of flow velocity and structural geometry) include frontal impact on the upstream face, drag along the sides,

3 FOUNDATION DESIGN LOADS

and suction on the downstream side. This manual assumes that the velocity of the floodwaters is constant (i.e., steady state flow).

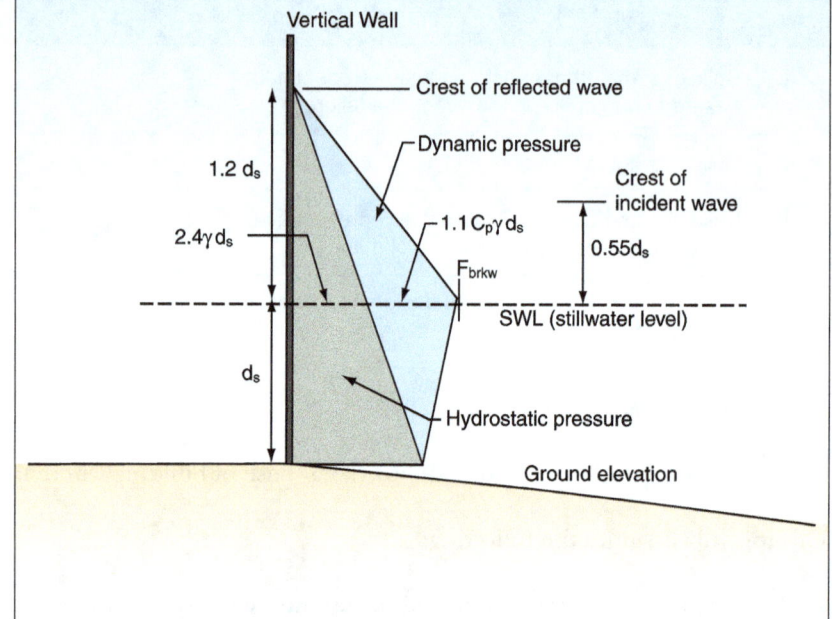

Figure 3-3. Normally incident breaking wave pressures against a vertical wall (space behind vertical wall is dry).

SOURCE: ASCE 7-05

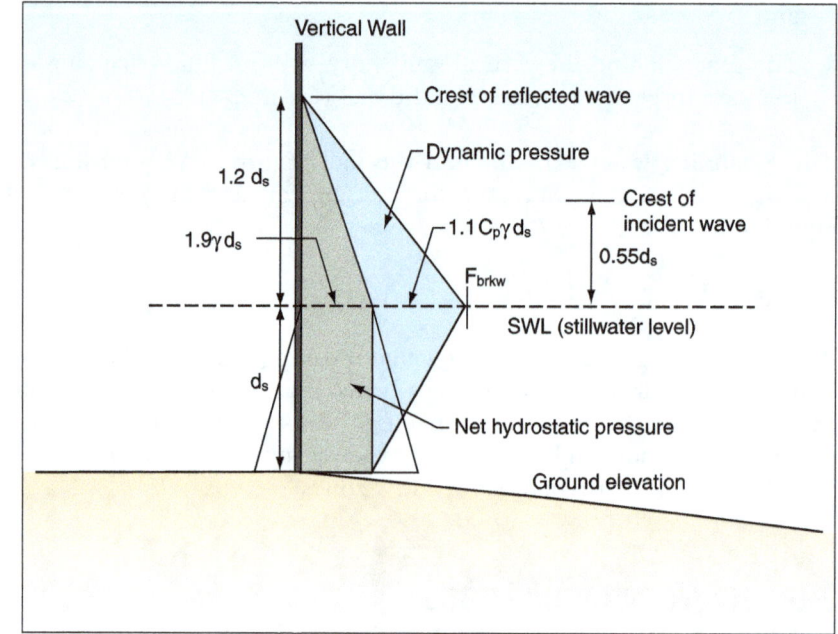

Figure 3-4. Normally incident breaking wave pressures against a vertical wall (stillwater level equal on both sides of wall).

SOURCE: ASCE 7-05

FOUNDATION DESIGN LOADS 3

One of the most difficult steps in quantifying loads imposed by moving water is determining the expected flood velocity. Refer to Section 3.2.4 for guidance concerning design flood velocities.

The following equation from FEMA 55 can be used to calculate the hydrodynamic load from flows with velocity greater than 10 ft/sec:

$$F_{dyn} = \tfrac{1}{2} C_d \rho V^2 A$$

Where

F_{dyn} = Hydrodynamic force (lb) acting at the stillwater mid-depth (halfway between the stillwater elevation and the eroded ground surface)

C_d = Drag coefficient (recommended values are 2.0 for square or rectangular piles and 1.2 for round piles)

ρ = Mass density of fluid (1.94 slugs/ft^3 for freshwater and 1.99 slugs/ft^3 for saltwater)

V = Velocity of water (ft/sec)

A = Surface area of obstruction normal to flow (ft^2)

Note that the use of this formula will provide the total force against a building of a given impacted surface area (A). Dividing the total force by either length or width would yield a force per unit length; dividing by "A" would yield a force per unit area.

The drag coefficient used in the previously stated equations is a function of the shape of the object around which flow is directed. If the object is something other than a round, square, or rectangular pile, the drag coefficient can be determined using Table 3-2.

Table 3-2. Drag Coefficient Based on Width to Depth Ratio

Width to Depth Ratio (w/d$_s$ or w/h)	Drag Coefficient (C$_d$)
1 to 12	1.25
13 to 20	1.30
21 to 32	1.40
33 to 40	1.50
41 to 80	1.75
81 to 120	1.80
>120	2.00

Note: "h" refers to the height of an object completely immersed in water.

SOURCE: *COASTAL CONSTRUCTION MANUAL* (FEMA 55)

3 FOUNDATION DESIGN LOADS

Flow around a building or building component will also create flow-perpendicular forces (lift forces). If the building component is rigid, lift forces can be assumed to be small. But if the building component is not rigid, lift forces can be greater than drag forces. The formula for lift force is similar to the formula for hydrodynamic force except that the drag coefficient (C_d) is replaced with the lift coefficient (C_l). For the purposes of this manual, the foundations of coastal residential buildings can be considered rigid, and hydrodynamic lift forces can therefore be ignored.

3.6 Debris Impact Loads

Debris or impact loads are imposed on a building by objects carried by moving water. The magnitude of these loads is very difficult to predict, yet some reasonable allowance must be made for them. The loads are influenced by where the building is located in the potential debris stream:

- Immediately adjacent to or downstream from another building
- Downstream from large floatable objects (e.g., exposed or minimally covered storage tanks)
- Among closely spaced buildings

The following equation to calculate the magnitude of impact load is provided in the *Commentary* of ASCE 7-05:

$$F_i = (\pi W V_b C_I C_O C_D C_B R_{max}) \div (2g\Delta t)$$

Where

F_i = Impact force acting at the stillwater level (lb)

π = 3.14

W = Weight of debris (lb), suggest using 1,000 if no site-specific information is available

V_b = Velocity of object (assume equal to velocity of water) (ft/sec)

C_I = Importance coefficient (see Table C5-1 of ASCE 7-05)

C_O = Orientation coefficient = 0.8

C_D = Depth coefficient (see Table C5-2 and Figure C5-1 of ASCE 7-05)

C_B = Blockage coefficient (see Table C5-3 and Figure C5-2 of ASCE 7-05)

R_{max} = Maximum response ratio for impulsive load (see Table C5-4 of ASCE 7-05)

g = Gravitational constant (32.2 ft/sec^2)

Δt = Duration of impact (sec)

When the C coefficients and R_{max} are set to 1.0, the above equation reduces to

$F_i = (\pi W V) \div (2g\Delta t)$

This equation is very similar to the equation provided in ASCE 7-98 and FEMA 55. The only difference is the $\pi/2$ term, which results from the half-sine form of the impulse load.

The following uncertainties must be quantified before the impact of debris loading on the building can be determined using the above equation:

- Size, shape, and weight (W) of the waterborne object
- Flood velocity (V)
- Velocity of the object compared to the flood velocity
- Portion of the building that will be struck and most vulnerable to collapsing
- Duration of the impact (t)

Once floodborne debris impact loads have been quantified, decisions must be made on how to apply them to the foundation and how to design foundation elements to resist them. For open foundations, the *Coastal Construction Manual* (FEMA 55) advises applying impact loading to a corner or critical column or pile concurrently with other flood loads (see FEMA 55, Table 11-6). For closed foundations (which are not recommended in Coastal A zones and are not allowed in V zones), FEMA 55 advises that the designer assume that one corner of the foundation will be destroyed by debris and recommends the foundation and the structure above be designed to contain redundancy to allow load redistribution to prevent collapse or localized failure. The following should be considered in determining debris impact loads:

Size, shape, and weight of the debris. It is recommended that, in the absence of information about the nature of the potential debris, a weight of 1,000 pounds be used for the debris weight (W). Objects of this weight could include portions of damaged buildings, utility poles, portions of previously embedded piles, and empty storage tanks.

Debris velocity. Flood velocity can be approximated by one of the equations discussed in Section 3.2.4. For the calculation of debris loads, the velocity of the waterborne object is assumed to be the same as the flood velocity. Note that, although this assumption may be accurate for small objects, it will overstate debris velocities for large objects (e.g., trees, logs, pier piles). The *Commentary* of ASCE 7-05 provides guidance on estimating debris velocities for large debris.

Portion of building to be struck. The object is assumed to be at or near the water surface level when it strikes the building. Therefore, the object is assumed to strike the building at the stillwater flood level.

Duration of impact. Uncertainty about the duration of impact (Δt) (the time from initial impact, through the maximum deflection caused by the impact, to the time the object leaves) is the most likely cause of error in the calculation of debris impact loads. ASCE 7-05 showed that measured impact duration (from initial impact to time of maximum force) from laboratory tests varied from 0.01 to 0.05 second. The ASCE 7-05 recommended value for Δt is 0.03 second.

>  **NOTE:** The method for determining debris impact loads in ASCE 7-05 was developed for riverine impact loads and has not been evaluated for coastal debris that may impact a building over several wave cycles. Although these impact loads are very large but of short duration, a structural engineer should be consulted to determine the structural response to the short load duration (0.03 second recommended).

3.7 Erosion and Localized Scour

Erosion is defined by Section 1-2 of ASCE 24-05 as the "wearing away of the land surface by detachment and movement of soil and rock fragments, during a flood or storm or over a period of years, through the action of wind, water, or other geological processes." Section 7.5 of FEMA 55 describes erosion as "the wearing or washing away of coastal lands." Since the exact configuration of the soil loss is important for foundation design purposes, a more specific definition is used in this document (see the text box above and Figure 3-5).

> **Erosion** refers to a general lowering of the ground surface over a wide area.
>
> **Scour** refers to a localized loss of soil, often around a foundation element.

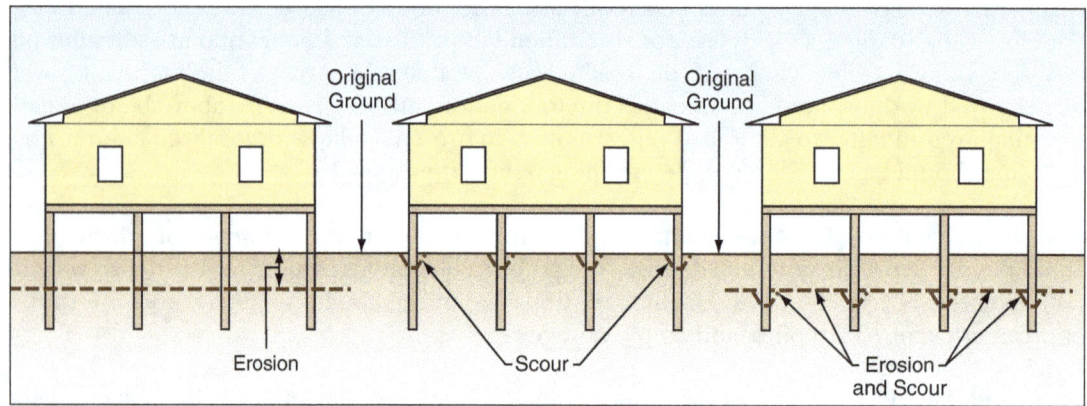

Figure 3-5. Distinguishing between coastal erosion and scour. A building may be subject to either or both, depending on the building location, soil characteristics, and flood conditions.

Waves and currents during coastal flood conditions are capable of creating turbulence around foundation elements and causing localized scour, and the moving floodwaters can cause generalized erosion. Determining potential for localized scour and generalized erosion is critical in designing coastal foundations to ensure that failure during and after flooding does not occur as a result of the loss in either bearing capacity or anchoring resistance around the posts, piles, piers, columns, footings, or walls. Localized scour and generalized erosion determinations will require knowledge of the flood depth, flow conditions, soil characteristics, and foundation type.

In some locations, soil at or below the ground surface can be resistant to localized scour, and scour depths calculated below will be excessive. In instances where the designer believes the soil at a site will be scour-resistant, a geotechnical engineer should be consulted before calculated scour depths are reduced.

3.7.1 Localized Scour Around Vertical Piles

The methods for calculating localized scour (S_{max}) in coastal areas have been largely based on empirical evidence gathered after storms. Much of the evidence gathered suggests that localized scour depths around piles and other thin vertical members are approximately equal to 1.0 to 1.5 times the pile diameter. Figure 3-6 illustrates localized scour at a pile, with and without a scour-resistant terminating stratum. Currently, there is no design guidance in ASCE 7-05 on how to calculate scour. FEMA 55 suggests that localized scour around a foundation element be calculated by the following equation:

$$S_{max} = 2.0a$$

Where

S_{max} = Maximum localized scour depth (ft)

a = Diameter of a round foundation element or the maximum diagonal cross-section dimension for a rectangular element (ft)

However, recent storms (e.g., Hurricane Ike, which struck the Texas coast in October 2008) have produced localized scour that exceeded the suggested depths. Because scour, coupled with erosion, can cause foundation systems to fail, a more conservative approach should be considered. Foundation systems should be analyzed for their ability to resist scour depths of 3 to 4 times pile diameters in addition to anticipated erosion levels. This guidance is more conservative than what has been recommended in FEMA 55, FEMA 499, and other publications.

NOTE: Resisting *higher bending moments* brought about by erosion and scour may necessitate a larger cross-section or decreased pile spacing (i.e., more piles) or, in some cases, use of a different pile material (e.g., concrete or steel instead of wood). Resisting *increased lateral flood loads* brought about by erosion (and possibly by linear scour) would necessitate a similar approach. However, designers should remember that increasing the number of piles or increasing the pile diameter will, in turn, also increase lateral flood loads on the foundation.

Resisting *increased unbraced lengths* brought about by erosion and scour will require deeper embedment of the foundation into the ground.

Erosion and scour can have several adverse impacts on coastal foundations:

■ Erosion and scour can reduce the embedment of the foundation into the soil, causing shallow foundations to collapse and making buildings on deep foundations more susceptible to settlement, lateral movement, or overturning from lateral loads.

3 FOUNDATION DESIGN LOADS

Figure 3-6.
Scour at vertical foundation member stopped by underlying scour-resistant stratum.

SOURCE: *COASTAL CONSTRUCTION MANUAL* (FEMA 55)

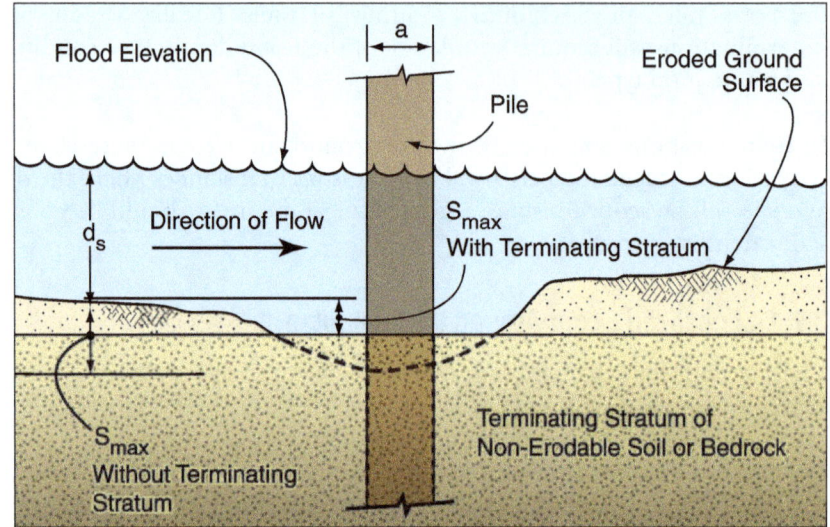

- Erosion and scour can increase the unbraced length of pile foundations, increase the bending moment to which they are subjected, and overstress piles.

- Erosion over a large area between a foundation and a flood source can expose the foundation to increased lateral flood loads (i.e., greater stillwater depths, possible higher wave heights, and higher flow velocities).

- Local scour around individual piles will not generally expose foundations to greater flood loads, but scour across a building site may do so.

To illustrate these points, calculations were made to examine the effects of erosion and scour on foundation design for a simple case – a 32-foot x 32-foot two-story home (10-foot story height), situated away from the shoreline and elevated 8 feet above grade on 25 square timber piles (spaced 8 feet apart), on medium dense sand. The home was subjected to a design wind event with a 130-mph (3-second gust speed) wind speed and a 4-foot stillwater depth above the uneroded grade, with storm surge and broken waves passing under the elevated building. Lateral wind and flood loads were calculated in accordance with ASCE. For simplicity, the piles were analyzed under lateral wind and flood loads only; dead, live, and wind uplift loads were neglected. If dead, live, and wind uplift loads were included in the analysis, deeper pile embedment and possibly larger piles may be needed.

Three different timber pile sizes (8-, 10-, and 12-inch square) were evaluated using pre-storm embedment depths of 10-, 15-, and 20-feet, and five different erosion and scour conditions (erosion = 0 or 1 foot; scour ranges from 2.0 times the pile diameter to 4.0 times the pile diameter). The results of the analysis are shown in Table 3-3. A shaded cell indicates the combination of pile size, pre-storm embedment, and erosion and scour would not provide the bending resistance and/or embedment required to resist the lateral loads imposed on them. The reason(s) for a foundation failure is indicated in each shaded cell, using "P" for pile failure due to bending and overstress within the pile and "E" for an embedment failure from the pile/soil interaction. An unshaded cell with "OK" indicates bending and foundation embedment criteria would *both* be satisfied by the particular pile size/pile embedment/erosion and scour combination.

Table 3-3. Example Foundation Adequacy Calculations for a Two-Story Home Supported on Square Timber Piles (and situated away from the shoreline, with storm surge passing under the home, a 130-mph wind zone, and soil is medium dense sand)

Pile Embedment Before Erosion and Scour	Erosion and Scour Conditions	Pile Diameter, a 8 inch	Pile Diameter, a 10 inch	Pile Diameter, a 12 inch
10 feet	Erosion = 0, Scour = 0	P, E	E	OK
	Erosion = 1 foot, Scour = 2.0 a	P, E	E	E
	Erosion = 1 foot, Scour = 2.5 a	P, E	E	E
	Erosion = 1 foot, Scour = 3.0 a	P, E	E	E
	Erosion = 1 foot, Scour = 4.0 a	P, E	P, E	E
15 feet	Erosion = 0, Scour = 0	P	OK	OK
	Erosion = 1 foot, Scour = 2.0 a	P	OK	OK
	Erosion = 1 foot, Scour = 2.5 a	P	OK	OK
	Erosion = 1 foot, Scour = 3.0 a	P	OK	OK
	Erosion = 1 foot, Scour = 4.0 a	P, E	P, E	E
20 feet	Erosion = 0, Scour = 0	P	OK	OK
	Erosion = 1 foot, Scour = 2.0 a	P	OK	OK
	Erosion = 1 foot, Scour = 2.5 a	P	OK	OK
	Erosion = 1 foot, Scour = 3.0 a	P	OK	OK
	Erosion = 1 foot, Scour = 4.0 a	P	P	OK

P = pile failure due to bending and overstress within the pile

E = embedment failure from the pile/soil interaction

OK = bending and foundation embedment criteria *both* satisfied by the particular pile size/pile embedment/erosion and scour combination

3 FOUNDATION DESIGN LOADS

A review of Table 3-3 shows several key points:

- Increasing pile embedment will not offset foundation inadequacy (bending failure) resulting from too small a pile cross-section or too weak a pile material.

- Increasing cross-section (or material strength) will not compensate for inadequate pile embedment.

- Given the building and foundation configuration used in the example, the 8-inch square pile is not strong enough to resist the lateral loads resulting from the 130-mph design wind speed under any of the erosion and scour conditions evaluated, even if there is no erosion or scour. Homes supported by 8-inch square timber piles, with embedment depths of 10 feet or less, will likely fail in large numbers when subjected to design or near design loads and conditions. Homes supported by deeper 8-inch piles may still be lost during a design event due to pile (bending failures).

- The 10-inch square pile is strong enough to resist bending under all but the most severe erosion and scour conditions analyzed.

- The 12-inch pile is the only pile size evaluated that satisfies bending requirements under all erosion and scour conditions analyzed. This pile works with 10 feet of embedment under the no erosion and scour condition. However, introducing as little as 1 foot of erosion and scour equal to twice the pile diameter was enough to render the foundation too shallow.

- Fifteen feet of pile embedment is adequate for both 10- and 12-inch piles subject to 1 foot of erosion and scour up to three times the pile diameter. However, when the scour is increased to four times the pile diameter (frequently observed following Hurricane Ike), 15 feet of embedment is inadequate for both piles. In general terms, approximately 11 feet of embedment is required *in this example home* to resist the loads and conditions after erosion and scour are imposed.

- The 12-inch pile with 20 feet of embedment was the only foundation that worked under all erosion and scour conditions analyzed. This pile design may be justified for the example home analyzed when expected erosion and scour conditions are unknown or uncertain.

 **CAUTION:** The results in Table 3-3 should not be used in lieu of building- and site-specific engineering analyses and foundation design. The table is intended for illustrative purposes only and is based upon certain assumptions and simplifications, and for the combinations of building characteristics, soil conditions, and wind and flood conditions described above. Registered design professionals should be consulted for foundation designs.

These analyses were based on only 1 foot of erosion, which historically is a relatively small amount. Many storms like Hurricanes Isabel, Ivan, and Ike caused much more extensive erosion. In some areas, these storms stripped away several feet of soil.

A foot of erosion is more damaging than a foot of scour. While scour reduces pile embedment and increases stresses within the pile, erosion reduces embedment, increases stresses, and, since it increases stillwater depths, it also increases the flood loads that the foundation must resist.

Table 3-3 suggests that increasing embedment beyond 15 feet is not necessary for 10- and 12-inch piles. This is only the case for relatively small amounts of erosion (like the 1 foot of erosion in the example). If erosion depths are greater, pile embedment must be increased.

3.7.2 Localized Scour Around Vertical Walls and Enclosures

Localized scour around vertical walls and enclosed areas (e.g., typical A zone construction) can be greater than that around vertical piles, and should be estimated using Table 3-4.

Table 3-4. Local Scour Depth as a Function of Soil Type

Soil Type	Expected Depth (% of d_s)
Loose sand	80
Dense sand	50
Soft silt	50
Stiff silt	25
Soft clay	25
Stiff clay	10

SOURCE: *COASTAL CONSTRUCTION MANUAL* (FEMA 55)

3.8 Flood Load Combinations

Load combinations (including those for flood loads) are given in ASCE 7-05, Sections 2.3.2 and 2.3.3 for strength design and Sections 2.4.1 and 2.4.2 for allowable stress design.

The basic load combinations are:

Allowable Stress Design

(1) $D + F$

(2) $D + H + F + L + T$

(3) $D + H + F + (L_r \text{ or } S \text{ or } R)$

(4) $D + H + F + 0.75(L + T) + 0.75(L_r \text{ or } S \text{ or } R)$

(5) $D + H + F + (W \text{ or } 0.7E)$

(6) $D + H + F + 0.75(W \text{ or } 0.7E) + 0.75L + 1.5F_a + 0.75(L_r \text{ or } S \text{ or } R)$

(7) $0.6D + W + H$

(8) $0.6D + 0.7E + H$

3 FOUNDATION DESIGN LOADS

Strength Design

(1) $1.4\,(D + F)$

(2) $1.2\,(D + F + T) + 1.6(L + H) + 0.5(L_r \text{ or } S \text{ or } R)$

(3) $1.2D + 1.6(L_r \text{ or } S \text{ or } R) + (L \text{ or } 0.8W)$

(4) $1.2D + 1.6W + L + 0.5(L_r \text{ or } S \text{ or } R)$

(5) $1.2D + 1.0E + L + 0.2S$

(6) $0.9D + 1.6W + 1.6H$

(7) $0.9D + 1.0E + 1.6H$

For structures located in V or Coastal A zones:

Allowable Stress Design

Load combinations 5, 6, and 7 shall be replaced with the following:

(5) $D + H + F + 1.5F_a + W$

(6) $D + H + F + 0.75W + 0.75L + 1.5F_a + 0.75(L_r \text{ or } S \text{ or } R)$

(7) $0.6D + W + H + 1.5F_a$

Strength Design

Load combinations 4 and 6 given in ASCE 7-05 Section 2.3.1 shall be replaced with the following:

(4) $1.2D + 1.6W + 2.0F_a + L + 0.5(L_r \text{ or } S \text{ or } R)$

(6) $0.9D + 1.6W + 2.0\,F_a + 1.6H$

Where

D = dead load

W = wind load

E = earthquake load

F_a = flood load

F = load due to fluids with well defined pressures and maximum heights

L = live load

L_r = roof live load

S = snow load

R = rain load

H = lateral earth pressure

Flood loads were included in the load combinations to account for the strong correlation between flood and winds in hurricane-prone regions that run along the Gulf of Mexico and the Atlantic Coast.

In non-Coastal A zones, for ASD, replace the $1.5F_a$ with $0.75F_a$ in load combinations 5, 6, and 7 given above. For strength design, replace coefficients W and F_a in equations 4 and 6 above with 0.8 and 1.0, respectively.

Designers should be aware that not all of the flood loads will act at certain locations or against certain building types. Table 3-5 provides guidance to designers for the calculation of appropriate flood loads in V zones and Coastal A zones (non-Coastal A zone flood load combinations are shown for comparison).

The floodplain management regulations enacted by communities that participate in the NFIP prohibit the construction of solid perimeter wall foundations in V zones, but allow such foundations in A zones. Therefore, the designer should assume that breaking waves will impact piles in V zones and walls in A zones. It is generally unrealistic to assume that impact loads will occur on all piles at the same time as breaking wave loads; therefore, this manual recommends that impact loads be evaluated for strategic locations such as a building corner.

Table 3-5. Selection of Flood Load Combinations for Design

Case 1	Pile or Open Foundation in V Zone (Required)
	F_{brkp} (on all piles) + F_i (on one corner or critical pile only)
	or
	F_{brkp} (on front row of piles only) + F_{dyn} (on all piles but front row) + F_i (on one corner or critical pile only)
Case 2	Pile or Open Foundation in Coastal A Zone (Recommended)
	F_{brkp} (on all piles) + F_i (on one corner or critical pile only)
	or
	F_{brkp} (on front row of piles only) + F_{dyn} (on all piles but front row) + F_i (on one corner or critical pile only)
Case 3	Solid (Wall) Foundation in Coastal A Zone (NOT Recommended)
	F_{brkp} (on walls facing shoreline, including hydrostatic component) + F_{dyn}; assume one corner is destroyed by debris, and design in redundancy
Case 4	Solid (Wall) Foundation in Non-Coastal A Zone (Shown for Comparison)
	$F_{sta} + F_{dyn}$

SOURCE: *COASTAL CONSTRUCTION MANUAL* (FEMA 55)

RECOMMENDED RESIDENTIAL CONSTRUCTION
FOR COASTAL AREAS

Building on Strong and Safe Foundations

4. Overview of Recommended Foundation Types and Construction for Coastal Areas

Chapters 1 through 3 discussed foundation design loads and calculations and how these issues can be influenced by coastal natural hazards. This chapter will tie all of these issues together with a discussion of foundation types and methods of constructing a foundation for a residential structure.

4 OVERVIEW OF RECOMMENDED FOUNDATION TYPES AND CONSTRUCTION FOR COASTAL AREAS

4.1 Critical Factors Affecting Foundation Design

Foundation construction types are dependent upon the following critical factors:

- Design wind speed
- Elevation height required by the BFE and local ordinances
- Flood zone
- Soil parameters

Soil parameters, like bearing capacities, shear coefficients, and subgrade moduli, are important in designing efficient and effective foundations. But, for the purpose of creating the standardized foundation concepts for use in a variety of sites, some soil parameters have been assumed (as in the case of bearing capacity for shallow foundations) and others have been stipulated (as those required to produce specific performance – as in the case for deep driven piles). Assumptions used in developing the foundations are listed in Appendix C, where stipulations on pile capacity are also listed in the individual drawings.

4.1.1 Wind Speed

The basic wind speed determines the wind velocity used in establishing wind loads for a building. It can also have a significant influence on the size and strength of foundations that support homes. Contemporary codes and standards like the IRC, IBC, and ASCE 7 specify basic wind speeds as 3-second gust wind speeds. Earlier versions of codes and standards specified wind speeds with different averaging periods. One example is the fastest mile wind speed that was specified in the 1988 (and earlier) versions of ASCE 7 and in pre-2000 versions of most model building codes.

The wind speed map shown in Figure 3-1 illustrates that the basic (3-second gust) wind speeds along most of the Gulf of Mexico, the Atlantic coast, and coastal Alaska range between 120 and 150 mph. The basic wind speed for most of the Pacific coast is 85 mph. Several areas in the Pacific Northwest are designated as special wind regions and wind speeds are dictated locally. The design wind speeds for many of the U.S. territories and protectorates are tabulated in ASCE 7.

To determine forces on the building and foundation, the wind speed is critical. Wind speed creates wind pressures that act upon the building. These pressures are proportional to the square of the wind speed, so a doubling of the wind speed increases the wind pressure by a factor of four. The pressure applied to an area of the building will develop forces that must be resisted. To transfer these forces from the building to the foundation, properly designed load paths are required. For the foundation to be properly designed, all forces including uplift, compression, and lateral must be taken into account.

Although wind loads are important in the design of a building, in coastal areas flood loads often have a much greater effect on the design of the foundation itself.

OVERVIEW OF RECOMMENDED FOUNDATION TYPES AND CONSTRUCTION FOR COASTAL AREAS 4

4.1.2 Elevation

The required height of the foundation depends on three factors: the DFE, the site elevation, and the flood zone. The flood zone dictates whether the lowest habitable finished floor must be placed at the DFE or, in the case of homes in the V zone, the bottom of the lowest horizontal member must be placed at the DFE. Figure 4-1 illustrates how the BFE, freeboard, erosion, and the ground elevation determine the foundation height required. While not required by the NFIP, V zone criteria are recommended for Coastal A zones. Stated mathematically:

$$H = DFE - G + Erosion$$

or

$$H = BFE - G + Erosion + Freeboard$$

Where

H = Required foundation height (in ft)

DFE = Design Flood Elevation

BFE = Base Flood Elevation

G = Non-eroded ground elevation

Erosion = Short-term plus long-term erosion

Freeboard = 2009 IRC required in SFHAs, locally adopted or owner desired freeboard

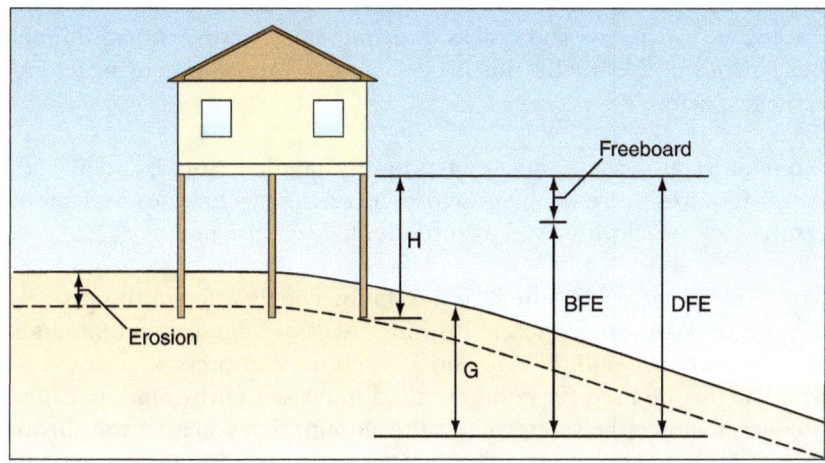

Figure 4-1.
The BFE, freeboard, erosion, and ground elevation determine the foundation height required.

The height to which a home should be elevated is one of the key factors in determining which pre-engineered foundation to use. Elevation height is dependent upon several factors, including the BFE, local ordinances requiring freeboard, and the desire of the homeowner to elevate the lowest horizontal structural member above the BFE (see also Chapter 2). This manual provides designs for closed foundations up to 8 feet above ground level and open foundations up to 15 feet above ground level. Custom designs can be developed for open and closed foundations to position the homes above those elevation levels. Foundations for homes

4 OVERVIEW OF RECOMMENDED FOUNDATION TYPES AND CONSTRUCTION FOR COASTAL AREAS

that need to be elevated higher than 15 feet should be designed by a licensed professional engineer.

4.1.3 Construction Materials

The use of flood-resistant materials below the BFE is also covered in FEMA NFIP Technical Bulletin 2, *Flood Damage-Resistant Materials Requirements for Buildings Located in Special Flood Hazard Areas in accordance with the National Flood Insurance Program* and FEMA 499, Fact Sheet No. 8 (see Appendix F). This manual will cover the materials used in masonry and concrete foundation construction, and field preservative treatment for wood.

4.1.3.1 Masonry Foundation Construction

The combination of high winds, moisture, and salt-laden air creates a damaging recipe for masonry construction. All three can penetrate the tiniest cracks or openings in the masonry joints. This can corrode reinforcement, weaken the bond between the mortar and the brick, and create fissures in the mortar. Moisture resistance is highly influenced by the quality of the materials and the workmanship.

4.1.3.2 Concrete Foundation Construction

Cast-in-place concrete elements in coastal environments should be constructed with 3 inches or more of concrete cover over the reinforcing bars. The concrete cover physically protects the reinforcing bars from corrosion. However, if salt water penetrates the concrete cover and reaches the reinforcing steel, the concrete alkalinity is reduced by the salt chloride, thereby corroding the steel. As the corrosion forms, it expands and cracks the concrete, allowing the additional entry of water and further corrosion. Eventually, this process weakens the concrete structural element and its load carrying capacity.

Alternatively, epoxy-coated reinforcing steel can be used if properly handled, stored, and placed. Epoxy-coated steel, however, requires more sophisticated construction techniques and more highly trained contractors than are usually involved with residential construction.

Concrete mix used in coastal areas must be designed for durability. The first step in this process is to start with the mix design. The American Concrete Institute (ACI) 318 manual recommends that a maximum water-cement ratio by weight of 0.40 and a minimum compressive strength of 4,000 pounds per square inch (psi) be used for concrete used in coastal environments. Since the amount of water in a concrete mix largely determines the amount that concrete will shrink and promote unwanted cracks, the water-cement ratio of the concrete mix is a critical parameter in promoting concrete durability. Adding more water to the mix to improve the workability increases the potential for cracking in the concrete and can severely affect its durability.

Another way to improve the durability of a concrete mix is with ideal mix proportions. Concrete mixes typically consist of a mixture of sand, aggregate, and cement. How these elements are proportioned is as critical as the water-cement ratio. The sand should be clean and free of contaminants. The aggregate should be washed and graded. The type of aggregate is also very important.

Recent research has shown that certain types of gravel do not promote a tight bond with the paste. The builder or contractor should consult expert advice prior to specifying the concrete mix.

Addition of admixtures such as pozzolans (fly ash) is recommended for concrete construction along the coast. Fly ash when introduced in concrete mix has benefits such as better workability and increased resistance to sulfates and chlorates, thus reducing corrosion from attacking the steel reinforcing.

4.1.3.3 Field Preservative Treatment for Wood Members

In order to properly connect the pile foundation to the floor framing system, making field cuts, notches, and boring holes are some of the activities associated with construction. Since pressure-preservative-treated piles, timbers, and lumber are used for many purposes in coastal construction, the interior, untreated parts of the wood are exposed to possible decay and infestation. Although treatments applied in the field are much less effective than factory treatments, the potential for decay can be minimized. The American Wood Preservers' Association (AWPA) *AWPA M4-08 Standard for the Care of Preservative-Treated Wood Products* (AWPA 2008) describes field treatment procedures and field cutting restrictions for poles, piles, and sawn lumber.

Field application of preservatives should always be done in accordance with instructions on the label. When detailed instructions are not provided, dip soaking for at least 3 minutes can be considered effective for field applications. When this is impractical, treatment may be done by thoroughly brushing or spraying the exposed area. It should be noted that the material is more absorptive at the end of a member, or end grains, than it is for the sides or side grains. To safeguard against decay in bored holes, the holes should be poured full of preservative. If the hole passes through a check (such as a shrinkage crack caused by drying), it will be necessary to brush the hole; otherwise, the preservative would run into the check instead of saturating the hole.

Waterborne arsenicals, pentachlorophenol, and creosote are unacceptable for field applications. Copper napthenate is the most widely used field treatment. Its deep green color may be objectionable, but the wood can be painted with alkyd paints in dark colors after extended drying. Zinc napthenate is a clear alternative to copper napthenate. However, it is not quite as effective in preventing insect infestation, and it should not be painted with latex paints. Tributyltin oxide (TBTO) is available, but should not be used in or near marine environments, because the leachates are toxic to aquatic organisms. Sodium borate is also available, but it does not readily penetrate dry wood and it rapidly leaches out when water is present. Therefore, sodium borate is not recommended.

4.1.4 Foundation Design Loads

To provide flexibility in the home designs, tension connections have been specified between the tops of all wood piles and the grade beams. Depending on the location of shear walls, shear wall openings, and the orientation of floor and roof framing, some wood piles may not experience tension forces. Design professionals can analyze the elevated structure to identify compression only piles to reduce construction costs. For foundation design and example calculations, see Appendix D.

4 OVERVIEW OF RECOMMENDED FOUNDATION TYPES AND CONSTRUCTION FOR COASTAL AREAS

Figure 4-2 illustrates design loads acting on a column. The reactions at the base of the elevated structure used in most of the foundation designs are presented in Tables 4-1a (one-story) and 4-1b (two-story). These reactions are the controlling forces for the range of building weights and dimensions listed in Appendix A and shown in Figure 2 of the Introduction. Design reactions have been included for the various design wind speeds and various building elevations above exterior grade. ASCE 7-05 load combination 4 ($D + 0.75L + 0.75L_r$) controls for gravity loading and load combination 7 controls for uplift and lateral loads. Load combination 7 is $0.6D + W + 0.75F_a$ in non-Coastal A zones and $0.6D + W + 1.5F_a$ in Coastal A and V zones. Refer to Section 3.8 for the list of flood load combinations.

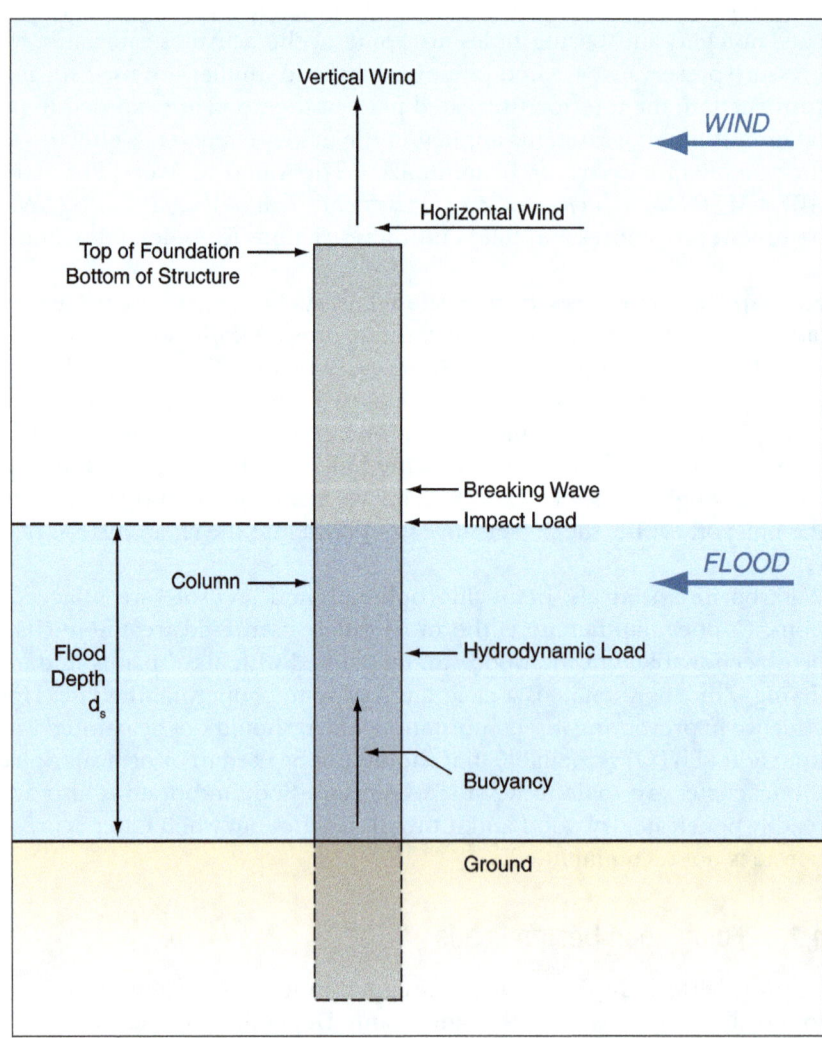

Figure 4-2.
Design loads acting on a column.

Loads on the foundation elements themselves are more difficult to tabulate because they depend on the foundation style (open or enclosed), foundation dimensions, and foundation height. Table 4-2 provides reactions for the 18-inch square columns used in most of the open foundation designs.

OVERVIEW OF RECOMMENDED FOUNDATION TYPES AND CONSTRUCTION FOR COASTAL AREAS 4

Table 4-1a. Design Perimeter Wall Reactions (lb/lf) for One-Story Elevated Homes (Note: Reactions are taken at the base of the elevated home/top of the foundation element.)

V H	120 mph Horiz	120 mph Vert	130 mph Horiz	130 mph Vert	140 mph Horiz	140 mph Vert	150 mph Horiz	150 mph Vert	(All V) Gravity
5 ft	770	-175	903	-259	1,048	-350	1,203	-448	1,172
6 ft	770	-175	903	-259	1,048	-350	1,203	-448	1,172
7 ft	770	-175	903	-259	1,048	-350	1,203	-448	1,172
8 ft	804	-202	944	-291	1,095	-388	1,257	-490	1,172
10 ft	804	-202	944	-291	1,095	-388	1,257	-490	1,172
12 ft	804	-202	944	-291	1,095	-388	1,257	-490	1,172
14 ft	832	-224	977	-317	1,133	-417	1,300	-525	1,172
15 ft	843	-226	989	-319	1,147	-419	1,317	-527	1,172

lb = pound
lf = linear foot
V = wind speed
H = height of foundation above grade
Horiz = horizontal
Vert = vertical

Table 4-1b. Design Perimeter Wall Reactions (lb/lf) for Two-Story Elevated Homes (Note: Reactions are taken at the base of the elevated home/top of the foundation element.)

V H	120 mph Horiz	120 mph Vert	130 mph Horiz	130 mph Vert	140 mph Horiz	140 mph Vert	150 mph Horiz	150 mph Vert	(All V) Gravity
5 ft	1,149	-145	1,348	-255	1,564	-374	1,795	-502	1,608
6 ft	1,149	-145	1,348	-255	1,564	-374	1,795	-502	1,608
7 ft	1,149	-145	1,348	-255	1,564	-374	1,795	-502	1,608
8 ft	1,182	-168	1,387	-282	1,609	-406	1,847	-539	1,608
10 ft	1,191	-171	1,397	-286	1,629	-410	1,860	-543	1,608
12 ft	1,191	-171	1,397	-286	1,620	-410	1,860	-543	1,608
14 ft	1,191	-171	1,397	-286	1,620	-410	1,860	-543	1,608
15 ft	1,210	-175	1,420	-291	1,647	-416	1,890	-550	1,608

b = pound
lf = linear foot
V = wind speed
H = height of foundation above grade
Horiz = horizontal
Vert = vertical

Table 4-2. Flood Forces (in pounds) on an 18-Inch Square Column

Flood Depth	Hydrodynamic	Breaking Wave	Impact	Buoyancy
5 ft	1,000	684	3,165	465
6 ft	1,440	985	3,476	577
7 ft	1,960	1,340	3,745	650

Table 4-2. Flood Forces (in pounds) on an 18-Inch Square Column (continued)

Flood Depth	Hydrodynamic	Breaking Wave	Impact	Buoyancy
8 ft	2,560	1,750	4,004	743
10 ft	4,001	2,735	4,476	939
12 ft	5,761	3,938	4,903	1,115
14 ft	7,841	5,360	5,296	1,300
15 ft	9,002	6,155	5,482	1,394

4.1.5 Foundation Design Loads and Analyses

Load analyses used to develop Case H foundations are similar to the analyses completed for the original FEMA 550 designs. Live loads used were those specified by the IRC and the original and augmented foundations were developed to support a range of dead loads. Wind and flood loads were calculated per ASCE 7, *Minimum Design Loads for Buildings and Other Structures* (the Case H design loads were calculated using ASCE 7-05; loads used in the original designs were calulated using ASCE 7-02, which are consistent with the 2005 edition). Design assumptions are listed in Appendix C.

Some noteworthy differences exist. Wind loads used in the original FEMA 550 designs were the worst case loads for a home that varied in width from 24 feet to 42 feet and in roof slope from 3:12 to 12:12. The foundation reactions for the original designs are listed in Table 4-1b. In the Case H designs, separate wind loads were determined based on the number of stories (one or two) and the building width (14 feet for the 3-bay designs, 28 feet for the 6-bay designs, and 42 feet for the 9-bay designs). The more precise matching of wind loads to building widths and heights provide greater design efficiencies.

Wind loads used to develop the Case H foundations are listed in Table 4-3.

Table 4-3. Wind Reactions Used to Develop Case H Foundations

	Two-Story								
	3-Bay			6-Bay			9-Bay		
	ww (p/lf)	lw (p/lf)	lat (p/lf)	ww (p/lf)	lw (p/lf)	lat (p/lf)	ww (p/lf)	lw (p/lf)	lat (p/lf)
120 mph	-710	320	420	-640	-50	510	-730	-270	610
130 mph	-840	380	490	-750	-60	600	-860	-310	710
140 mph	-970	440	570	-870	-70	700	-1,000	-360	830
150 mph	-1,100	500	650	-1,000	-80	800	-1,140	-410	950

Table 4-3. Wind Reactions Used to Develop Case H Foundations (continued)

	One-Story							
	3-Bay			6-Bay			9-Bay		
	ww (p/lf)	lw (p/lf)	lat (p/lf)	ww (p/lf)	lw (p/lf)	lat (p/lf)	ww (p/lf)	lw (p/lf)	lat (p/lf)
120 mph	-340	-20	240	-440	-200	320	-580	-340	410
130 mph	-400	-20	280	-510	-230	380	-680	-400	480
140 mph	-470	-30	320	-600	-270	440	-790	-460	560
150 mph	-540	-30	370	-690	-310	500	-900	-530	640

ww = vertical forces on windward edge of foundation
lw = vertical forces on leeward edge of foundation
lat = horizontal forces on windward and leeward edges of foundation
1. (+) loads act upward; (-) pressures act downward.
2. Lateral loads are applied to both windward and leeward foundation elements.

To account for shear panel reactions from segmented shear walls, the analyses of foundations supporting one-story homes included 6.72 kip quarter span point loads for the 3-bay design (point loads were applied at mid-span for the 6- and 9-bay models, Figure 4-3). The loads correspond to 10-foot tall wood framed shear panels constructed with 7/16-inch blocked wood structural panels fastened with 8d common nails 6 inches on center (o.c.). Foundations supporting two-story homes were analyzed with 13.44 kip shear panel reactions or twice that of the one-story home. The foundations will also support homes constructed with perforated shear walls.

Another difference in design methodology was required due to the nature of structural frames. In the original designs, the concrete columns were considered statically determinant and analyzed as such. The structural frames created by the concrete grade beams, concrete columns, and elevated beams, however, are not statically determinant and computer modeling was warranted. To analyze the frame action developed by those structural elements, computer models using RISA© structural software were created. Design loads were applied to the frames and critical shears and moments were tabulated for the grade beams, columns, and elevated beams. Critical axial forces were also tabulated for the columns.

Tables 4-4 through 4-9 summarize the critical shears, moments, and axial loads of the computer models used to develop the Case H foundations.

Figure 4-3.
Shear panel reactions for the 3- and 6-bay models. Reactions for the 9-bay model were similar to those of the 6-bay.

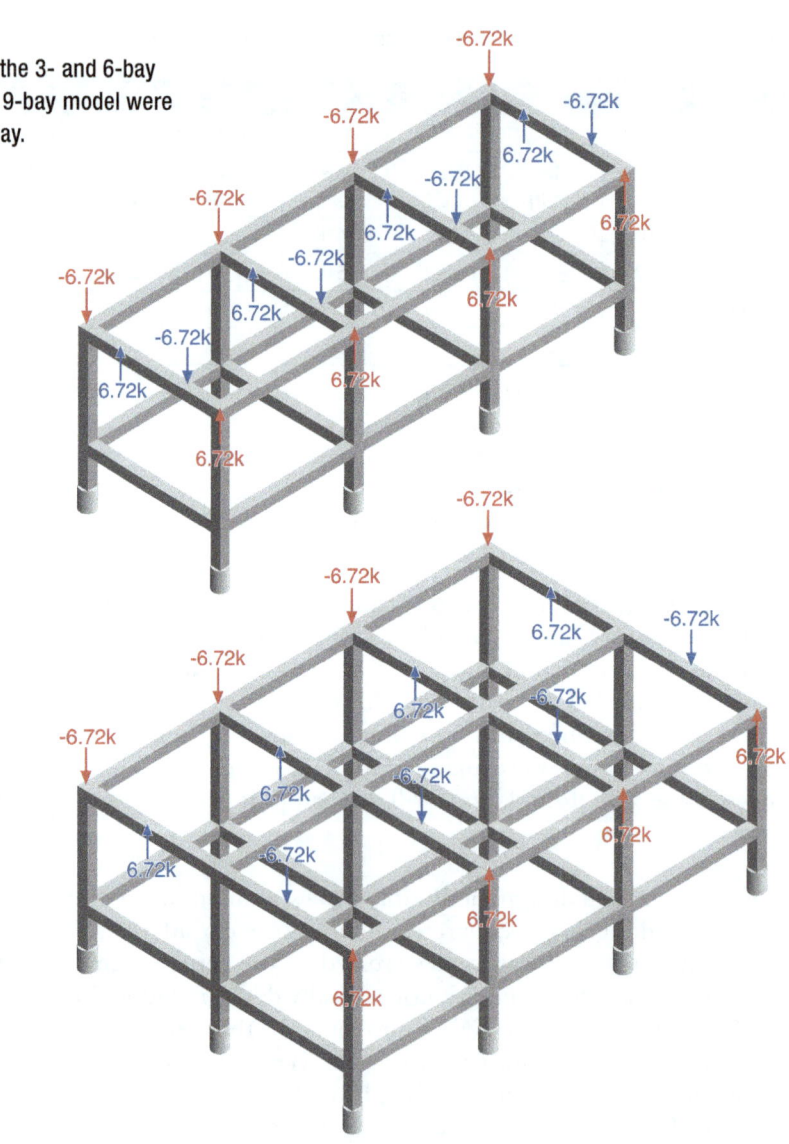

Table 4-4. Design Moments (K-ft), Axial Loads (in kips), and Shears (in kips) for 10-Foot Tall 3-Bay Foundations

10-Foot Foundation	3-Bay One-Story				3-Bay Two-Story			
	120 mph	130 mph	140 mph	150 mph	120 mph	130 mph	140 mph	150 mph
Column Moment +	24	25	27	30	33	38	44	51
Column Moment -	37	40	43	47	51	57	63	69
Column Shear Bottom	9	9	10	10	11	12	13	14
Column Shear Top	5	6	6	8	8	9	11	12

OVERVIEW OF RECOMMENDED FOUNDATION TYPES AND CONSTRUCTION FOR COASTAL AREAS 4

Table 4-4. Design Moments (K-ft), Axial Loads (in kips), and Shears (in kips) for 10-Foot Tall 3-Bay Foundations (continued)

10-Foot Foundation	3-Bay One-Story				3-Bay Two-Story			
	120 mph	130 mph	140 mph	150 mph	120 mph	130 mph	140 mph	150 mph
Axial Maximum	24	24	24	25	46	48	49	50
Axial Minimum	8	8	7	7	1.6	1.2	0.5	0.3
Elevated Beam Moment +	22	24	26	29	39	43	46	47
Elevated Beam Moment -	21	23	25	27	41	44	48	51
Elevated Beam Shear at Column	3	3	3	3	4	4	3	3
Elevated Beam Shear at Mid-Span	7	7	8	8	13	14	15	15
Grade Beam Moment +	30	30	30	31	41	42	44	46
Grade Beam Moment -	18	18	19	20	26	28	30	32
Grade Beam Shear at Column	9	9	9	9	10	10	10	10
Grade Beam Shear at Mid-Span	4	4	4	4	5	5	6	6

1. (+) loads act upward; (-) pressures act downward.

Table 4-5. Design Moments (K-ft), Axial Loads (in kips), and Shears (in kips) for 15-Foot Tall 3-Bay Foundations

15-Foot Foundation	3-Bay One-Story				3-Bay Two-Story			
	120 mph	130 mph	140 mph	150 mph	120 mph	130 mph	140 mph	150 mph
Column Moment +	53	57	62	67	73	81	90	98
Column Moment -	79	84	88	95	104	112	122	132
Column Shear Bottom	13	13	14	15	15	17	18	19
Column Shear Top	6	6	7	8	9	10	11	12
Axial Maximum	27	27	27	28	46	48	50	53
Axial Minimum	7	7	6	6	12	1	-0.3	-1.6
Elevated Beam Moment +	51	55	59	65	67	79	87	95
Elevated Beam Moment -	39	46	50	55	63	68	78	86
Elevated Beam Shear at Column	4	5	5	6	3	4	5	6
Elevated Beam Shear at Mid-Span	11	12	13	13	19	20	21	22
Grade Beam Moment +	30	30	30	29	36	37	39	41
Grade Beam Moment -	21	21	21	21	26	27	29	31

4 OVERVIEW OF RECOMMENDED FOUNDATION TYPES AND CONSTRUCTION FOR COASTAL AREAS

Table 4-5. Design Moments (K-ft), Axial Loads (in kips), and Shears (in kips) for 15-Foot Tall 3-Bay Foundations (continued)

15-Foot Foundation	3-Bay One-Story				3-Bay Two-Story			
	120 mph	130 mph	140 mph	150 mph	120 mph	130 mph	140 mph	150 mph
Grade Beam Shear at Column	9	9	9	9	10	10	9	10
Grade Beam Shear at Mid-Span	4	4	4	4	5	5	4	5

1. (+) loads act upward; (-) pressures act downward.

Table 4-6. Design Moments (K-ft), Axial Loads (in kips), and Shears (in kips) for 10-Foot Tall 6-Bay Foundations

10-Foot Foundation	6-Bay One-Story				6-Bay Two-Story			
	120 mph	130 mph	140 mph	150 mph	120 mph	130 mph	140 mph	150 mph
Column Moment +	56	59	62	65	77	84	90	97
Column Moment -	74	78	81	85	69	75	81	88
Column Shear Bottom	19	20	21	21	17	18	19	20
Column Shear Top	8	9	10	11	14	15	17	19
Axial Maximum	32	32	30	30	47	47	46	46
Axial Minimum	8	8	7	6	10	9	9	8
Elevated Beam Moment +	29	31	32	33	42	44	46	48
Elevated Beam Moment -	23	23	23	23	42	42	43	43
Elevated Beam Shear at Column	9	9	9	9	14	14	14	15
Elevated Beam Shear at Mid-Span	3	3	3	3	7	7	7	8
Grade Beam Moment +	71	73	77	80	60	64	69	74
Grade Beam Moment -	53	75	79	83	55	61	67	73
Grade Beam Shear at Column	13	13	14	14	11	12	12	14
Grade Beam Shear at Mid-Span	10	11	11	11	8	9	10	11

1. (+) loads act upward; (-) pressures act downward.

Table 4-7. Design Moments (K-ft), Axial Loads (in kips), and Shears (in kips) for 15-Foot Tall 6-Bay Foundations

15 Foot Foundation	6-Bay One-Story				6-Bay Two-Story			
	120 mph	130 mph	140 mph	150 mph	120 mph	130 mph	140 mph	150 mph
Column Moment +	81	87	92	98	118	127	137	140
Column Moment -	100	105	111	118	129	139	150	157
Column Shear Bottom	15	16	17	18	20	21	22	24

OVERVIEW OF RECOMMENDED FOUNDATION TYPES AND CONSTRUCTION FOR COASTAL AREAS   4

Table 4-7. Design Moments (K-ft), Axial Loads (in kips), and Shears (in kips) for 15-Foot Tall 6-Bay Foundations (continued)

15 Foot Foundation	6-Bay One-Story				6-Bay Two-Story			
	120 mph	130 mph	140 mph	150 mph	120 mph	130 mph	140 mph	150 mph
Column Shear Top	9	10	11	11	13	14	16	18
Axial Maximum	37	37	32	33	48	48	48	45
Axial Minimum	-1	-2	-4	-5	1	-1	-1	-1
Elevated Beam Moment +	43	46	49	52	62	66	71	75
Elevated Beam Moment -	23	22	24	27	44	44	44	41
Elevated Beam Shear at Column	11	11	12	12	16	16	17	18
Elevated Beam Shear at Mid-Span	2	2	1	2	4	5	3	3
Grade Beam Moment +	99	104	109	114	119	127	136	140
Grade Beam Moment -	107	114	121	129	128	138	150	155
Grade Beam Shear at Column	19	20	21	21	22	23	20	19
Grade Beam Shear at Mid-Span	15	16	16	18	18	19	18	15

1. (+) loads act upward; (-) pressures act downward.

Table 4-8. Design Moments (K-ft), Axial Loads (in kips), and Shears (in kips) for 10-Foot Tall 9-Bay Foundations

10-Foot Foundation	9-Bay One-Story				9-Bay Two-Story			
	120 mph	130 mph	140 mph	150 mph	120 mph	130 mph	140 mph	150 mph
Column Moment +	21	23	26	29	24	28	32	36
Column Moment -	40	43	47	50	52	56	61	63
Column Shear Bottom	9	9	10	10	10	11	12	13
Column Shear Top	4	5	6	7	6	7	9	10
Axial Maximum	32				47			
Axial Minimum	8				10			
Elevated Beam Moment +	29	29	29	29	47	47	47	47
Elevated Beam Moment -	15	15	15	15	25	25	25	25
Elevated Beam Shear at Column	11	11	11	11	18	18	18	18
Elevated Beam Shear at Mid-Span	12	12	12	12	20	20	20	20
Grade Beam Moment +	45	48	51	54	55	58	63	68
Grade Beam Moment -	44	47	50	54	56	60	66	71

4 OVERVIEW OF RECOMMENDED FOUNDATION TYPES AND CONSTRUCTION FOR COASTAL AREAS

Table 4-8. Design Moments (K-ft), Axial Loads (in kips), and Shears (in kips) for 10-Foot Tall 9-Bay Foundations (continued)

10-Foot Foundation	9-Bay One-Story				9-Bay Two-Story			
	120 mph	130 mph	140 mph	150 mph	120 mph	130 mph	140 mph	150 mph
Grade Beam Shear at Column	9	10	10	10	10	11	12	13
Grade Beam Shear at Mid-Span	7	7	8	7	7	8	8	9

1. (+) loads act upward; (-) pressures act downward.

Table 4-9. Design Moments (K-ft), Axial Loads (in kips), and Shears (in kips) for 15-Foot Tall 9-Bay Foundations

15 Foot Foundation	9-Bay One-Story				9-Bay Two-Story			
	120 mph	130 mph	140 mph	150 mph	120 mph	130 mph	140 mph	150 mph
Column Moment +	49	54	59	98	51	65	73	81
Column Moment -	80	86	92	118	100	106	115	124
Column Shear Bottom	12	13	14	18	14	15	16	17
Column Shear Top	6	6	7	11	7	8	10	11
Axial Maximum	34	39	34	34	49	49	49	49
Axial Minimum	6	6	6	6	9	8	8	8
Elevated Beam Moment +	29	29	29	29	48	48	48	48
Elevated Beam Moment -	16	16	16	13	26	26	26	26
Elevated Beam Shear at Column	15	12	12	12	20	20	20	20
Elevated Beam Shear at Mid-Span	2	1	1	1	2	2	2	2
Grade Beam Moment +	95	100	105	114	111	117	125	133
Grade Beam Moment -	103	109	116	128	123	132	141	151
Grade Beam Shear at Column	19	19	20	22	21	22	23	25
Grade Beam Shear at Mid-Span	14	15	16	18	17	18	19	21

1. (+) loads act upward; (-) pressures act downward.

4.2 Recommended Foundation Types for Coastal Areas

Table 4-10 provides six open (deep and shallow) foundation types and two closed foundations discussed in this manual. Appendix A provides the foundation design drawings for the cases specified.

OVERVIEW OF RECOMMENDED FOUNDATION TYPES AND CONSTRUCTION FOR COASTAL AREAS 4

Table 4-10. Recommended Foundation Types Based on Zone

Foundation		Case	V Zones	A Zones in Coastal Areas	
				Coastal A Zone	A Zone
Open Foundation (deep)	Braced timber pile	A	✔	✔	✔
	Steel pipe pile with concrete column and grade beam	B	✔	✔	✔
	Timber pile with concrete column and grade beam	C	✔	✔	✔
	Timber pile with concrete grade and elevated beams and concrete columns	H	✔	✔	✔
Open Foundation (shallow)	Concrete column and grade beam	D	NR	✔	✔
	Concrete column and grade beam with integral slab	G	NR	✔	✔
Closed Foundation (shallow)	Reinforced masonry – crawlspace	E	✘	NR	✔
	Reinforced masonry – stem wall	F	✘	NR	✔

✔ = Acceptable

NR = Not Recommended

✘ = Not Permitted

The foundation designs contained in this manual are based on soils having a bearing capacity of 1,500 pounds per square foot (psf). The 1,500-psf bearing capacity value corresponds to the presumptive value contained in Section 1806 of the 2009 IBC. The presumptive bearing capacity is for clay, sandy clay, silty clay, clayey silt, and sandy silt (CL, ML, MH, and CH soils).

The size of the perimeter footings and grade beams are generally not controlled by bearing capacity (uplift and lateral loads typically control footing size and grade beam dimensions). Refining the designs for soils with greater bearing capacities may not significantly reduce construction costs. However, the size of the interior pad footings for the crawlspace foundation (Table 4-10, Case E) depends greatly on the soil's bearing capacity. Design refinements can reduce footing sizes in areas where soils have greater bearing capacities. The following discussion of the foundation designs listed in Table 4-10 is also presented in Appendix A. Figures 4-4 through 4-10 are based on Appendix A.

4.2.1 Open/Deep Foundation: Timber Pile (Case A)

This pre-engineered, timber pile foundation uses conventional, tapered, treated piles and steel rod bracing to support the elevated structure. No concrete, masonry, or reinforcing steel is needed (see Figure 4-4). Often called a "stilt" foundation, the driven timber pile system is

4 OVERVIEW OF RECOMMENDED FOUNDATION TYPES AND CONSTRUCTION FOR COASTAL AREAS

suitable for moderate elevations if the homebuilder prefers to minimize the number of different construction trades used. Once the piles are driven, the wood guides and floor system are attached to the piles; the remainder of the home is constructed off the floor platform.

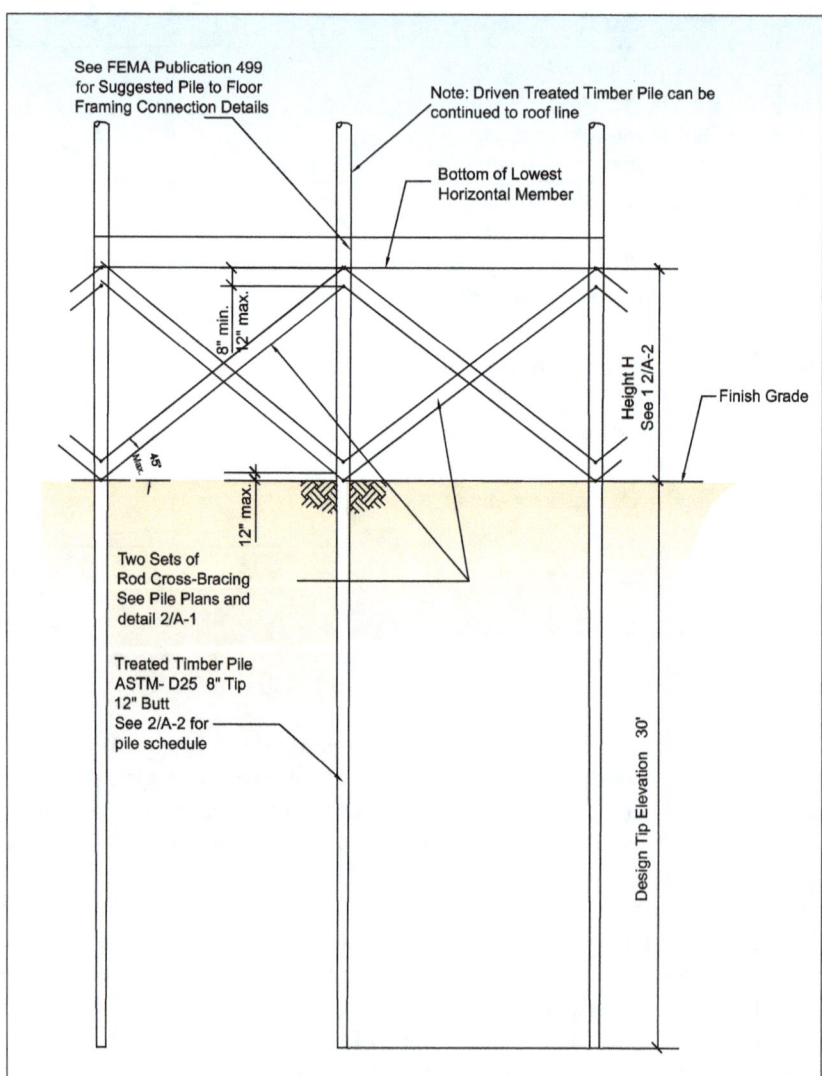

Figure 4-4. Profile of Case A foundation type (see Appendix A for additional drawings).

The recommended design for Case A that is presented in this manual accommodates home elevations up to 10 feet above grade. With customized designs and longer piles, the designs can be modified to achieve higher elevations. However, elevations greater than 10 feet will likely be prevented by pile availability, the pile strength required to resist lateral forces, and the pile embedment required to resist erosion and scour. A construction approach that can improve performance is to extend the piles above the first floor diaphragm to the second floor or roof diaphragm. Doing so allows the foundation and the elevated home to function more like a single, integrated structural frame. Extending the piles stiffens the structure, reduces stresses in the piles, and reduces lateral deflections. Post disaster assessments of pile supported homes indicate that extending piles in this fashion improves survivability. Licensed professional engineers should be consulted to analyze the pile foundations and design the appropriate connections.

One drawback of the timber pile system is the exposure of the piles to floodborne debris. During a hurricane event, individual piles can be damaged or destroyed by large, floating debris. With the home in place, damaged piles are difficult to replace. Two separate ways of addressing this potential problem is to use piles with a diameter larger than is called for in the foundation design or to use a greater number of piles to increase structural redundancy.

4.2.2 Open/Deep Foundation: Steel Pipe Pile with Concrete Column and Grade Beam (Case B)

This foundation incorporates open-ended steel pipe piles; this style is somewhat unique to the Gulf Coast region where the prevalence of steel pipe piles used to support oil platforms has created local sources for these piles. Like treated wood piles, steel pipe piles are driven but have the advantage of greater bending strength and load carrying capacity (see Figure 4-5). The open steel pipe pile foundation is resistant to the effects of erosion and scour. The grade beam can be undermined by scour without compromising the entire foundation system.

The number of piles required depends on local soil conditions. Like other soil dependent foundation designs, consideration should be given to performing soil tests on the site so the foundation design can be optimized. With guidance from engineers, the open-ended steel pipe pile foundation can be designed for higher elevations. Additional piles can be driven for increased resistance to lateral forces, and columns can be made larger and stronger to resist the increased bending moments that occur where the columns join the grade beam. Because only a certain amount of steel can be installed to a given cross-section of concrete before the column sizes and the flood loads become unmanageable, a maximum elevation of 15 feet exists for the use of this type of foundation.

4.2.3 Open/Deep Foundation: Timber Pile with Concrete Column and Grade Beam (Case C)

This foundation is similar to the steel pipe pile with concrete column and grade beam foundation (Case B). Elevations as high as 15 feet can be achieved for wind speeds up to 150 mph for both one- and two-story structures. However, because wood piles have a lower strength to resist the loads than steel piles, approximately twice as many timber piles are needed to resist loads imposed on the home and the exposed portions of the foundation (Figure 4-6).

While treated to resist rot and damage from insects, wood piles may become vulnerable to damage from wood destroying organisms in areas where they are not constantly submerged by groundwater. If constantly submerged, there is not enough oxygen to sustain fungal growth and insect colonies; if only periodically submerged, the piles can have moisture levels and oxygen

4 OVERVIEW OF RECOMMENDED FOUNDATION TYPES AND CONSTRUCTION FOR COASTAL AREAS

levels sufficient to sustain wood destroying organisms. Consultation with local design professionals in the area familiar with the use and performance of driven treated wood piles will help quantify this potential risk. Grade beams can be constructed at greater depths or alternative pile materials can be selected if wood destroying organism damage is a major concern.

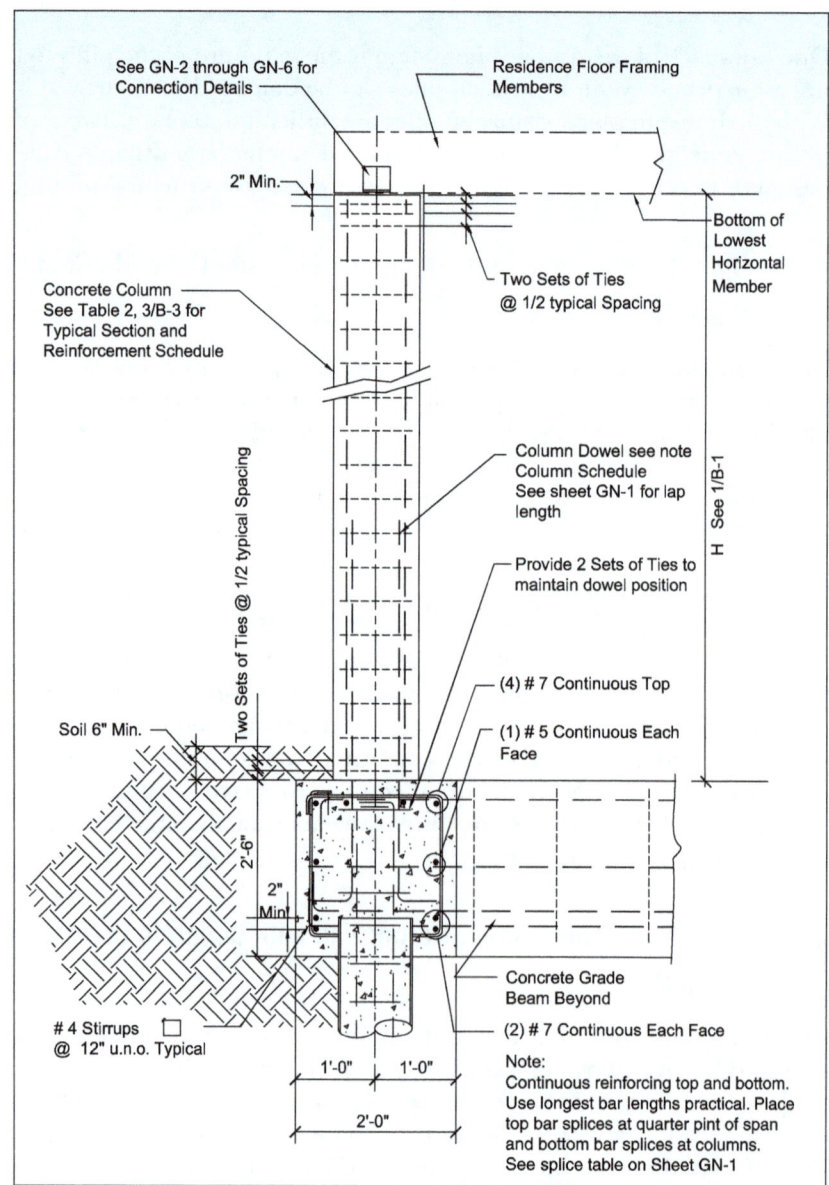

Figure 4-5. Profile of Case B foundation type (see Appendix A for additional drawings).

OVERVIEW OF RECOMMENDED FOUNDATION TYPES AND CONSTRUCTION FOR COASTAL AREAS 4

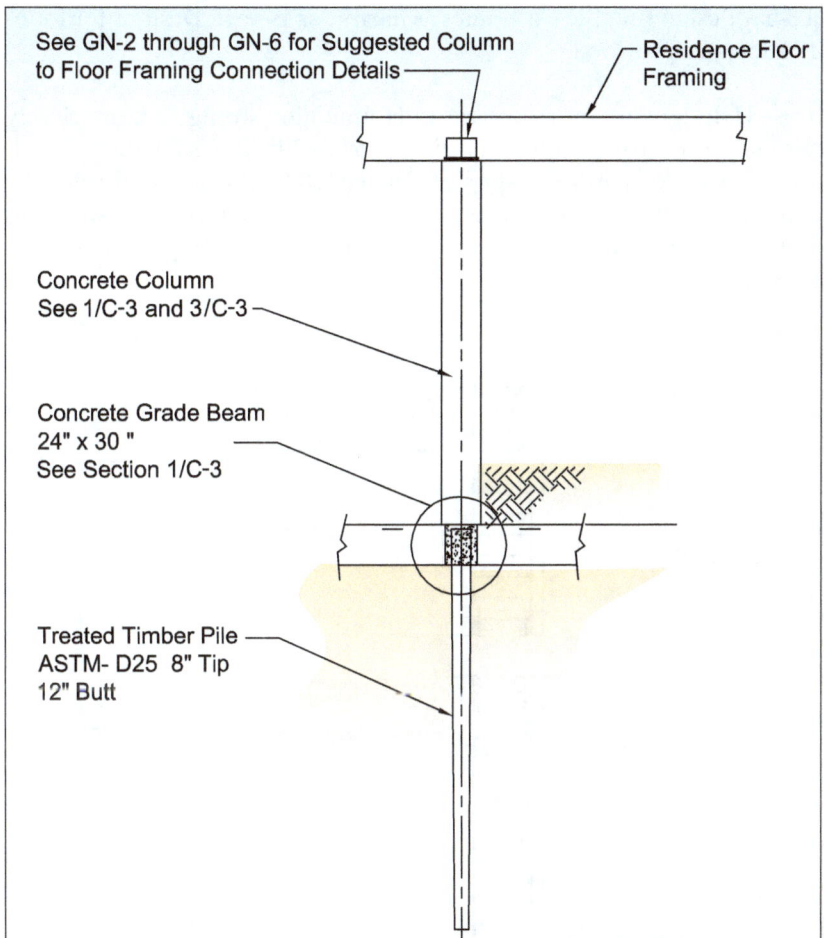

Figure 4-6.
Profile of Case C foundation type (see Appendix A for additional drawings).

4.2.4 Open/Deep Foundation: Timber Pile with Concrete Grade and Elevated Beams and Concrete Columns (Case H)

Case H foundation designs augment designs contained in the first edition of FEMA 550. They incorporate elevated reinforced beams into the V zone timber pile foundation design. The elevated beams provide two important benefits:

1. The elevated beams, columns, and grade beams function as structural frames that resist lateral loads. The frame action allows smaller concrete columns to be used. Smaller columns reduce the flood loads imposed on the foundation and provide more efficient designs.

2. The elevated beams provide attachment points for homes constructed per prescriptive codes and standards like ANSI/AF&PA *Wood Frame Construction Manual*, American Iron and Steel Institute (AISI) *Standard for Cold-Formed Steel Framing – Prescriptive Method for One and Two Family Dwellings*, SSTD10-99 *Standard for Hurricane Resistant Residential Construction*, and ICC-600 *Standard for Residential Construction in High Wind Regions*.

4 OVERVIEW OF RECOMMENDED FOUNDATION TYPES AND CONSTRUCTION FOR COASTAL AREAS

Case H designs include a 3-bay design suitable for homes as narrow as 14 feet. Designs for foundation heights of 10 and 15 feet are provided.

As previously stated, the Case H designs are more precise and foundation strengths more closely match design loads. In addition, the structural frame action provided by the grade beams, columns, and elevated beams allow smaller columns to be constructed. One drawback of the design, however, is that constructing elevated concrete beams is more complicated than constructing grade beams and reinforced columns; therefore, more knowledgeable and experienced contractors would be needed (Figure 4-7).

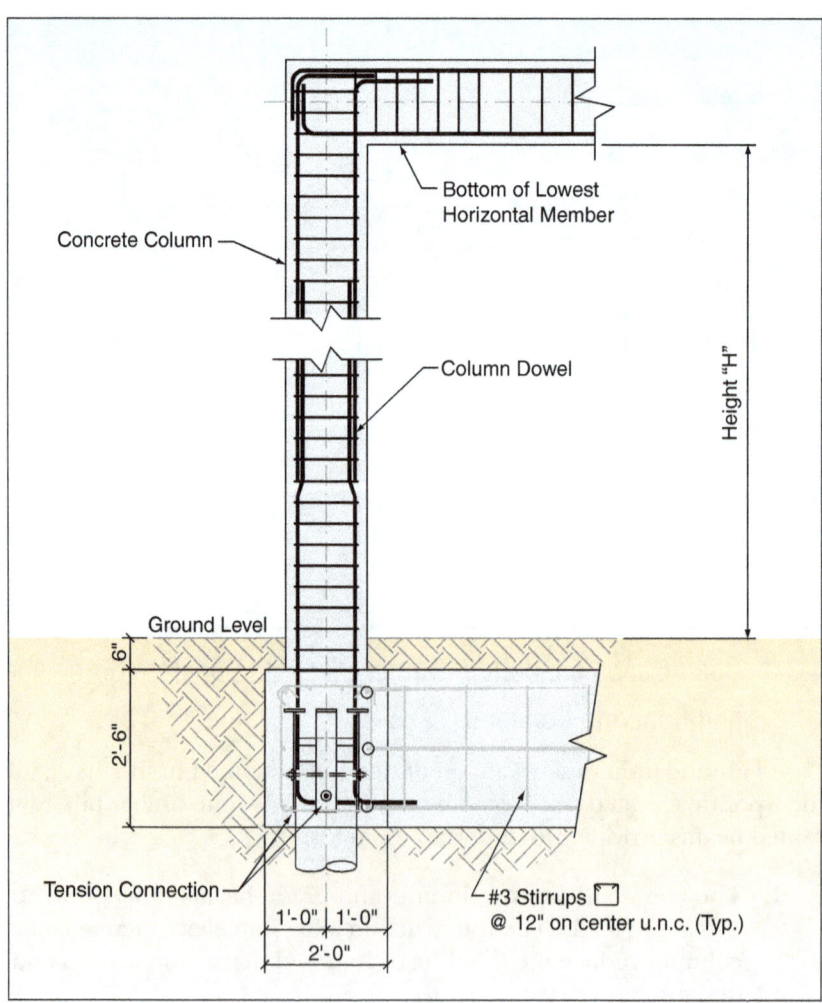

Figure 4-7.
Profile of Case H foundation type (see Appendix A for additional drawings).

Sections were designed with an emphasis on strength, ductility, and constructability. To simplify detailing and construction, axial and shear reinforcement were made consistent through each section. Section properties throughout each member were selected to resist the maximum forces (positive and negative moments, axial loads, and shears) that exist within the elements (grade beams, columns, or elevated beams). To simplify forming, the dimensions of the elevated beams matched the columns into which they frame. The designs were based on a belief that the potential increase in concrete costs would be more than offset by the savings in labor costs in constructing simple forms. Design professionals using the guidance contained in this manual may find it beneficial to vary from this approach.

The approach used to design and detail the Case H foundation system was to develop easily scalable column and beam systems. The size allows for the use of variable amounts of steel while keeping rebar congestion to a minimum. A 16-inch wide system to allow for the economical use of structural form panels with minimal waste was selected. The consistent column and elevated beam sizes also allow for reuse of the concrete forms between columns and beams. The member size and aspect ratio (member shape factor) allow for high lateral capacities.

The continuous bars in the beams provide a tie around the entire structure, imparting redundancy should an element fail, as well as the ability for the system to bridge over failed elements below. This redundancy improves the foundation performance, especially with impact from floodborne debris that exceeds design loads.

4.2.5 Open/Shallow Foundation: Concrete Column and Grade Beam with Slabs (Cases D and G)

These open foundation types make use of a rigid mat to resist lateral forces and overturning moments. Frictional resistance between the grade beams and the supporting soils resist lateral loads while the weight of the grade beam and the above grade columns resist uplift. Case G (foundation with slab) contains additional reinforcement to tie the on-grade slab to the grade beams to provide additional weight to resist uplift (Figure 4-8). With the integral slab, elevations up to 15 feet above grade are achievable. Without the slab (as for Case D), the designs as detailed are limited to 10-foot elevations (Figure 4-9).

Unlike the deep driven pile foundations, both shallow grade beam foundation styles can be undermined by erosion and scour if exposed to waves and high flow velocities. Neither style of foundation should be used where anticipated erosion or scour would expose the grade beam.

4.2.6 Closed/Shallow Foundation: Reinforced Masonry – Crawlspace (Case E)

The reinforced masonry with crawlspace type of foundation utilizes conventional construction similar to foundations used outside of SFHAs. Footings are cast-in-place reinforced concrete; walls are constructed with reinforced masonry (Figure 4-10). The foundation designs presented in Appendix A permit elevated homes to be raised to 8 feet. Higher elevations are achievable with larger or more closely spaced reinforcing steel or with walls constructed with thicker masonry.

The required strength of a masonry wall is determined by breaking wave loads for wall heights 3 feet or less, by non-breaking waves and hydrodynamic loads for taller walls, and by uplift for all walls. Perimeter footing sizes are controlled by uplift and must be relatively large for short foundation walls. The weight of taller walls contributes to uplift resistance and allows for

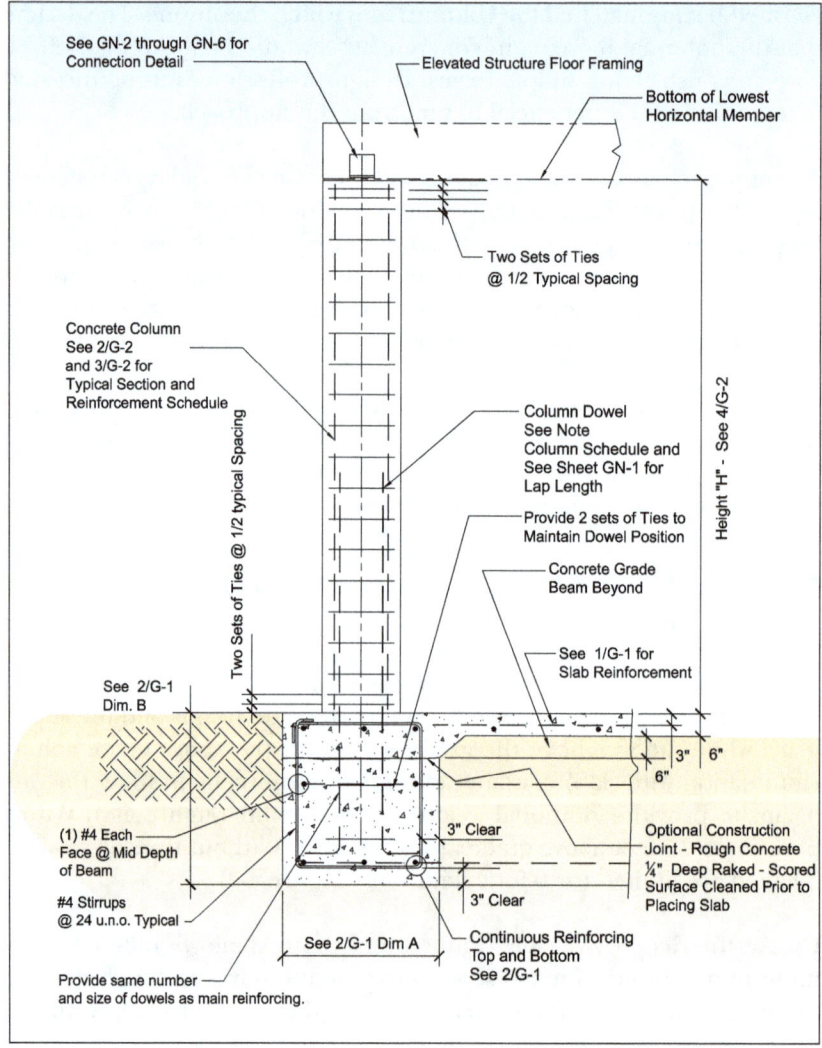

Figure 4-8. Profile of Case G foundation type (see Appendix A for additional drawings).

smaller perimeter footings. Solid grouting of perimeter walls is recommended for additional weight and improved resistance to water infiltration.

Interior footing sizes are controlled by gravity loads and by the bearing capacity of the supporting soils. Since the foundation designs are based on relatively low bearing capacities, obtaining soils tests for the building site may allow the interior footing sizes to be reduced.

The crawlspace foundation walls incorporate NFIP required flood vents, which must allow floodwaters to flow into the crawlspace. In doing so, hydrostatic, hydrodynamic, and breaking wave loads are reduced. Crawlspace foundations are vulnerable to scour and flood forces and should not be used in Coastal A zones; the NFIP prohibits their use in V zones.

OVERVIEW OF RECOMMENDED FOUNDATION TYPES AND CONSTRUCTION FOR COASTAL AREAS 4

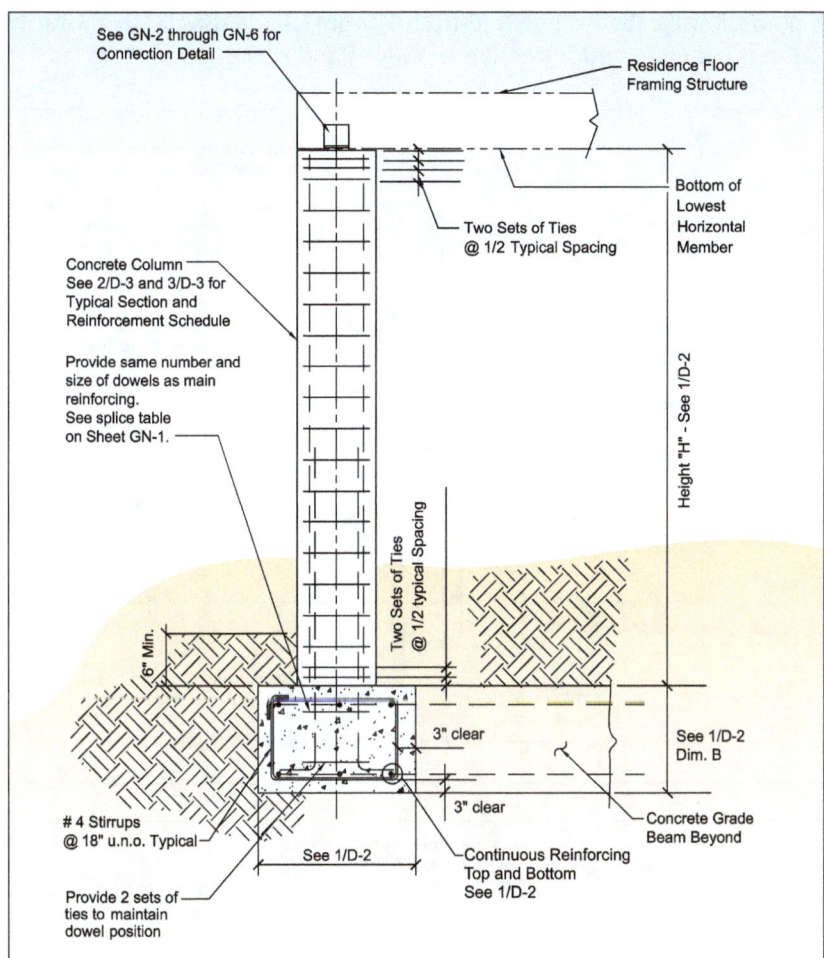

Figure 4-9. Profile of Case D foundation type (see Appendix A for additional drawings).

4.2.7 Closed/Shallow Foundation: Reinforced Masonry – Stem Wall (Case F)

The reinforced masonry stem walls (commonly referred to as chain walls in portions of the Gulf Coast) type of foundation also utilizes conventional construction to contain fill that supports the floor slab. They are constructed with hollow masonry block with grouted and reinforced cells (Figure 4-11). Full grouting is recommended to provide increased weight, resist uplift, and improve longevity of the foundation.

The amount and size of the reinforcement are controlled primarily by the lateral forces created by the retained soils and by surcharge loading from the floor slab and imposed live loads. Because the retained soils can be exposed to long duration flooding, loads from saturated soils should be considered in the analyses. The lateral forces on stem walls can be relatively high and even short cantilevered stem walls (those not laterally supported by the floor slab) need to be heavily reinforced. Tying the top of the stem walls into the floor slab provides lateral support for the walls and significantly reduces reinforcement requirements. Because backfill needs to be placed before the slab is poured, walls that will be tied to the floor slab need to be temporarily braced when the foundation is backfilled until the slab is poured and cured.

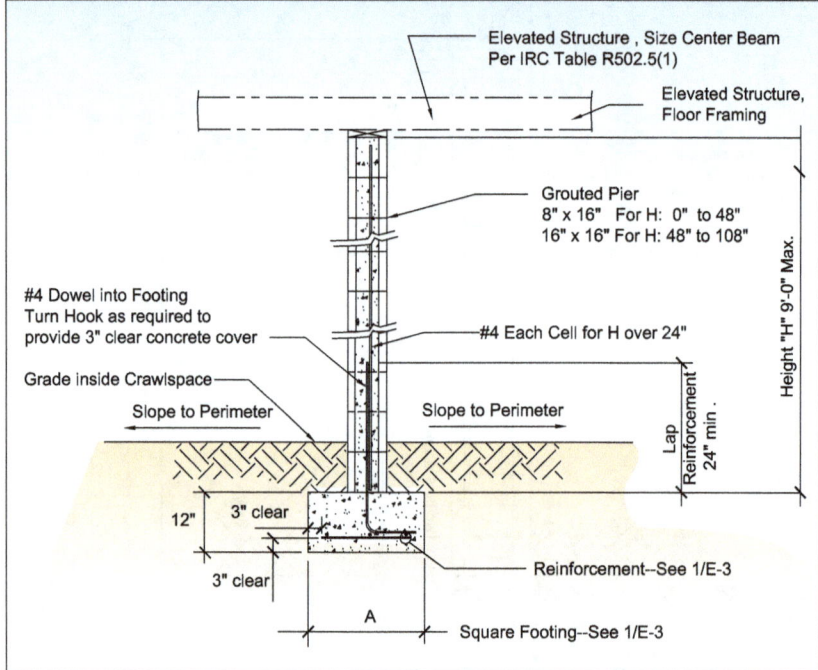

Figure 4-10. Profile of Case F foundation type (see Appendix A for additional drawings).

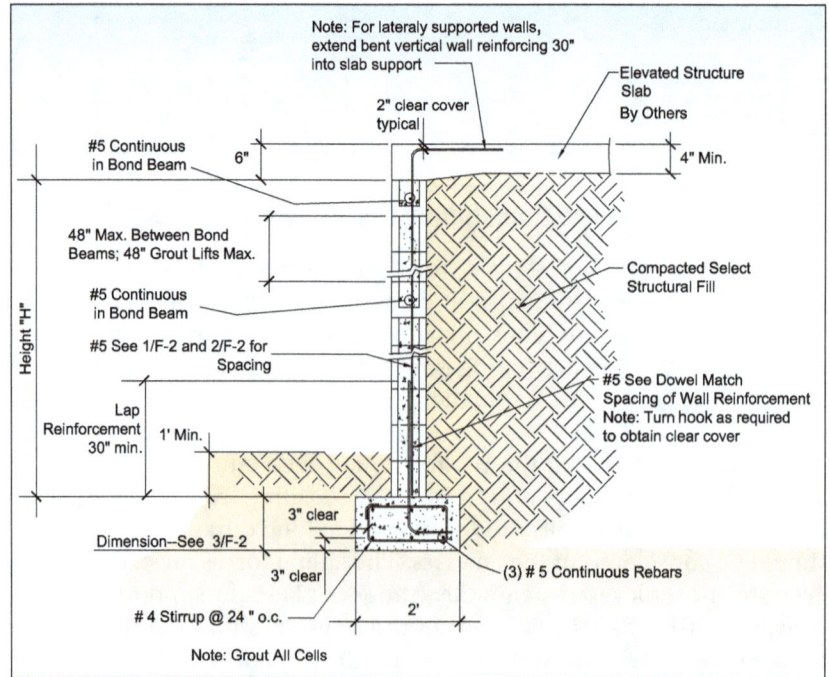

Figure 4-11. Profile of Case E foundation type (see Appendix A for additional drawings).

 **NOTE:** Stem wall foundations are vulnerable to scour and should not be used in Coastal A zones without a deep footing. The NFIP prohibits the use of this foundation type in V zones.

RECOMMENDED RESIDENTIAL CONSTRUCTION FOR COASTAL AREAS

Building on Strong and Safe Foundations

5. Foundation Selection

This chapter provides foundation designs, along with the use of the drawings in Appendix A, to assist the homebuilder, contractor, and local engineering professional in developing a safe and strong foundation. Foundation design types, foundation design considerations, cost estimating, and details on how to use this manual are presented.

5.1 Foundation Design Types

The homebuilder, contractor, and local engineering professional can utilize the designs in this chapter and Appendix A to construct residential foundations in coastal areas. The selection of appropriate foundation designs for the construction of residences is dependent upon the coastal zone, wind speed, and elevation requirements, all of which have been discussed in the previous chapters. The following types of foundation designs are presented in this manual:

Open/Deep Foundations

- Braced timber pile (Case A)
- Steel pipe pile with concrete column and grade beam (Case B)
- Timber pile with concrete column and grade beam (Case C)
- Timber pile with concrete grade and elevated beams and concrete columns (Case H)

Open/Shallow Foundations

- Concrete column and grade beam (Case D)
- Concrete column and grade beam with slab (Case G)

Closed/Shallow Foundations

- Reinforced masonry – crawlspace (Case E)
- Reinforced masonry – stem wall (Case F)

Each of these foundation types designed for coastal areas have advantages and disadvantages that must be taken into account. Modifications to the details and drawings might be needed to incorporate specific home footprints, elevation heights, and wind speeds to a given foundation type. Consultation with a licensed professional engineer is encouraged prior to beginning construction.

The foundation designs and materials specified in this document are based on principles and practices used by structural engineering professionals with years of coastal construction experience. This manual has been prepared to make the information easy to understand.

Guidance on the use of the foundation designs recommended herein is provided in Appendix B. Examples of how the foundation designs can be used with some of the homes in the publication *A Pattern Book for Gulf Coast Neighborhoods* are presented in Appendix B. Design drawings for each of the foundation types are presented in Appendix A, and any assumptions used in these designs are in Appendix C.

5.2 Foundation Design Considerations

The foundation designs proposed are suitable for homes with dimensions, weights, and roof pitches within certain ranges of values. A licensed professional engineer should confirm the appropriateness of the foundation design of homes with dimensions, weights, or roof pitches that fall outside of those defined ranges.

Most of the foundation designs are based on a 14-foot wide (maximum) by 24-foot deep (minimum) "module" (Figure 5-1). From this basic building block, foundations for specific homes can be developed. For example, if a 30-foot deep by 42-foot wide home is to be constructed, the foundation can be designed around three 14-foot wide by 30-foot deep sections. If a 24-foot deep by 50-foot wide home is desired, four 12.5-foot wide by 24-foot deep sections can be used. If a 22-foot deep home is desired, the foundation designs presented here should only be used after a licensed professional engineer determines that they are appropriate since the shallow depth of the building falls outside the range of assumptions used in the design.

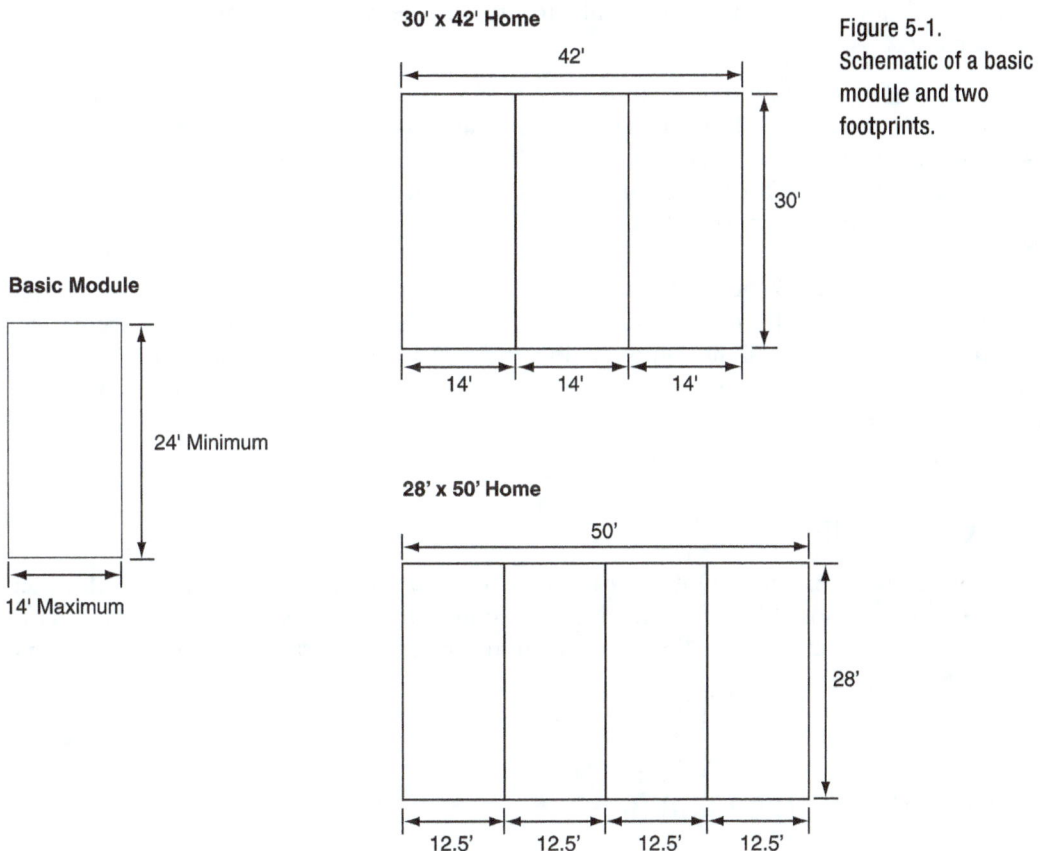

Figure 5-1. Schematic of a basic module and two footprints.

The licensed professional engineer should also consider the following:

- **Local soil conditions.** The pile foundations have been developed for relatively soft subsurface soils. For driven treated lumber piles, the presumptive allowable working load values of 7 tons per pile gravity, 4.65 tons per pile uplift, and 2 tons per pile lateral were used. For steel pipe piles, the presumptive allowable working load piles were greater (10 tons per pile for gravity loading, 6.7 tons per pile for uplift, and 4 tons per pile for lateral loading). Soil testing on the site should also be considered to validate the assumptions made.

  In some areas of the coastal U.S. (e.g., portions of Louisiana), soils may exist that will not provide the presumptive pile values. In those areas, aspects of the FEMA 550 deep

5 FOUNDATION SELECTION

foundation designs are still valid, but geotechnical engineers will need to be involved in portions of the design to determine required pile parameters. In poor soils, additional piles may need to be installed or long piles may need to be driven; however, the portions of the designs from the grade beams upward should remain valid.

The FEMA 550 shallow foundations are based on a presumptive bearing capacity of 1,500 psf. This value is consistent with the presumptive bearing capacity of Section 1806 of the 2009 IBC for clay, sandy clays, clayey silts, and sandy silts (CL, ML, MH, and CH soils). In areas where soils will not provide this presumed bearing capacity, the shallow FEMA 550 designs should not be used until their ability to support the required loads can be confirmed by design professionals.

- **Building weight.** The foundations have been designed to resist uplift forces resulting from a relatively light structure. If the actual home is heavier (e.g., from the use of concrete composite siding or steel framing), it may be cost-effective to reanalyze and redesign the footings. This is particularly true for a home that doesn't need to be elevated more than several feet or has short foundation walls that can help resist uplift.

- **Footprint complexity.** By necessity, the foundations have been designed for relatively simple rectangular footprints. If the actual footprint of the home is relatively complex, the engineer may need to consider torsional wind loading, differential movement among the "modules" that make up the home, concentrated loading in the home's floor and roof diaphragms, and shear wall placement.

5.3 Cost Estimating

Cost information that homebuilders can use to estimate the cost of installing the foundation systems proposed in this manual are presented in Appendix E. These cost estimates are based on May 2006 prices from information provided by local contractors for the First Edition of this manual.

5.4 How to Use This Manual

The rest of this chapter is designed to provide the user with step by step procedures for the information contained in this manual.

1. **Determine location of the dwelling on a general map.** Identify the location relative to key features such as highways and bodies of water. An accurate location is essential for using flood and wind speed maps in subsequent steps of the design process.

2. **Determine location of dwelling on the appropriate FIRM**

- Determine the flood insurance risk zone from the FIRM (Select V zone, Coastal A zone, non-Coastal A zone, or other). Refer to FEMA 258, *Guide to Flood Maps, How to Use Flood Maps to Determine Flood Risk for a Property*, for instructions.

FOUNDATION SELECTION 5

- Determine the BFE or the interim Advisory Base Flood Elevations (ABFEs) for the location from the FIRM. If the dwelling is outside of flood-prone areas, flood loads do not need to be considered.

    FIRM Panel No. _____

    Flood Insurance Risk Zone _____

    Base Flood Elevation (BFE) or
    Advisory Base Flood Elevation (ABFE) _____

3. Identify the local building code. Several states and municipalities in coastal areas are adopting new building codes to govern residential construction. This manual assumes that the IRC governs the design and construction requirements.

    County/Parish/City _____

    Building Code _____

    Building Code Date _____

4. Identify the local freeboard requirements and DFE. Using either the local building codes, local floodplain ordinances, data obtained from local building officials, or personal preferences (only if greater than minimum requirements), determine the minimum freeboard above the BFE or ABFE. The DFE is the sum of the BFE or ABFE and freeboard values.

    Base Flood Elevation (BFE) or
    Advisory Base Flood Elevation (ABFE) _____

    Freeboard   + _____

    Design Flood Elevation (DFE) _____

5. Determine the required design wind velocity. The 2006 and 2009 IRCs reference ASCE 7-05 as the source of the wind speed information.

    Design Wind Velocity _____

    Wind Exposure Category _____

6. Establish the topographic elevation of the building site and the dwelling. Elevations can be obtained from official topographic maps published by the National Geodetic Survey (NGS) and/or as established or confirmed by a surveyor.

RECOMMENDED RESIDENTIAL CONSTRUCTION FOR COASTAL AREAS

5 FOUNDATION SELECTION

- If the dwelling and its surrounding site are above the DFE, no flood forces need to be considered.

- If the desired topographic elevation is below the DFE, the dwelling must be elevated above the BFE or ABFE.

Source of Topo Elevation	_____
Topo Elevation (Site)	_____

7. Determine the height of the base of the dwelling above grade. Subtract the lowest ground elevation at the building from the lowest elevation of the structure (i.e., bottom of lowest horizontal structural member).

Design Flood Elevation (DFE)	_____
Topo Elevation	_____
Elevation Dimension	_____

8. Determine the general soil classification for the site. For shallow foundations, confirm that the soils on site have a minimum bearing capacity of 1,500 psf. If soils lack that minimum capacity, contact a geotechnical engineer and/or a structural engineer to confirm that the FEMA 550 foundation solutions are appropriate.

For deep foundations, confirm that the presumptive pile capacities (for gravity loads, uplift loads, and lateral loads) are achievable. If soils present on site will not support the presumed pile capacities, contact a geotechnical engineer and/or a structural engineer to determine appropriate pile plans.

Soil Classification	_____

9. Estimate erosion and scour. Estimate accumulated erosion and episodic scour over the life of the structure. Use accumulated erosion to determine eroded grade elevation and use accumulated erosion and episodic scour to determine the foundation depth required to ensure shallow foundations will not be undermined.

10. Determine the type of foundation to be used to support the structure. Depending on the location of the dwelling, design wind speed, and local soil conditions documented above, select the desired or required type of foundation. Note that more than one solution may be possible. Refer to Chapter 4 for the potential foundation designs that can be used within the flood zones determined from the FIRM maps. Drawings in Appendix A illustrate the construction details for each of the foundations. Refer to the drawings for further direction and information about the needs for each type of unit.

11. Evaluate alternate foundation type selections. The choice of foundation type may be on the basis of least cost or to provide a personal choice, functional, or aesthetic need at the site. Refer to Appendix E for guidance on preparing cost estimates. Functional needs such as provisions for parking, storage, or other non-habitable uses for the area beneath the living space should be considered in the selection of the foundation design. Aesthetic or architectural issues (i.e., appearance) also must be included in the evaluation process. Guidance for the architectural design considerations can be obtained from *A Pattern Book for Gulf Coast Neighborhoods* by the Mississippi Governor's Commission on Recovery, Rebuilding and Renewal (see Appendix B) and from many other sources.

As part of the final analysis, it is strongly recommended that the selection and evaluation process be coordinated with or reviewed by knowledgeable contractors or design professionals to arrive at the best solution to fulfill all of the regulatory and functional needs for the construction.

12. Select the foundation design. If the home's dimensions, height, roof pitch, and weight are within the ranges used to develop these designs, the foundation designs can be used "as is." However, if the proposed structure has dimensions, height, roof pitch, or weights that fall outside of the range of values used, a licensed professional engineer should be consulted. The materials presented in the appendices should help reduce the engineering effort needed to develop a custom design. Figure 5-2 is a foundation selection decision tree for determining which foundation design to use based on the requirements of the home. Tables 5-1a and 5-1b show which foundation design cases can be used for one- and two-story homes, respectively, based on height of elevation and wind velocity.

Because the designs are good for a range of buildings, they will be conservative for some applications. A licensed professional engineer will be able to provide value engineering and may produce a more efficient design that reduces construction costs.

5.5 Design Examples

The foundation designs were developed to allow a "modular approach" for developing foundation plans. In this approach, individual rectangular foundation components can be assembled into non-rectangular building footprints (see Figures 5-3 through 5-5). Appendix D provides detailed calculations and analysis for open and closed foundation designs. There are, however, a few rules that must be followed when assembling the modules:

1. **The eave-to-ridge dimension of the roof is limited to 23 feet.** The upper limit on roof height is to limit the lateral forces to those used in developing the designs.

2. **Roof slopes shall not be shallower than 3:12 or steeper than 12:12.** For a 12:12 roof pitch, this corresponds to a 42-foot deep home with a 2-foot eave overhang.

5 FOUNDATION SELECTION

Figure 5-2.
Foundation selection decision tree.

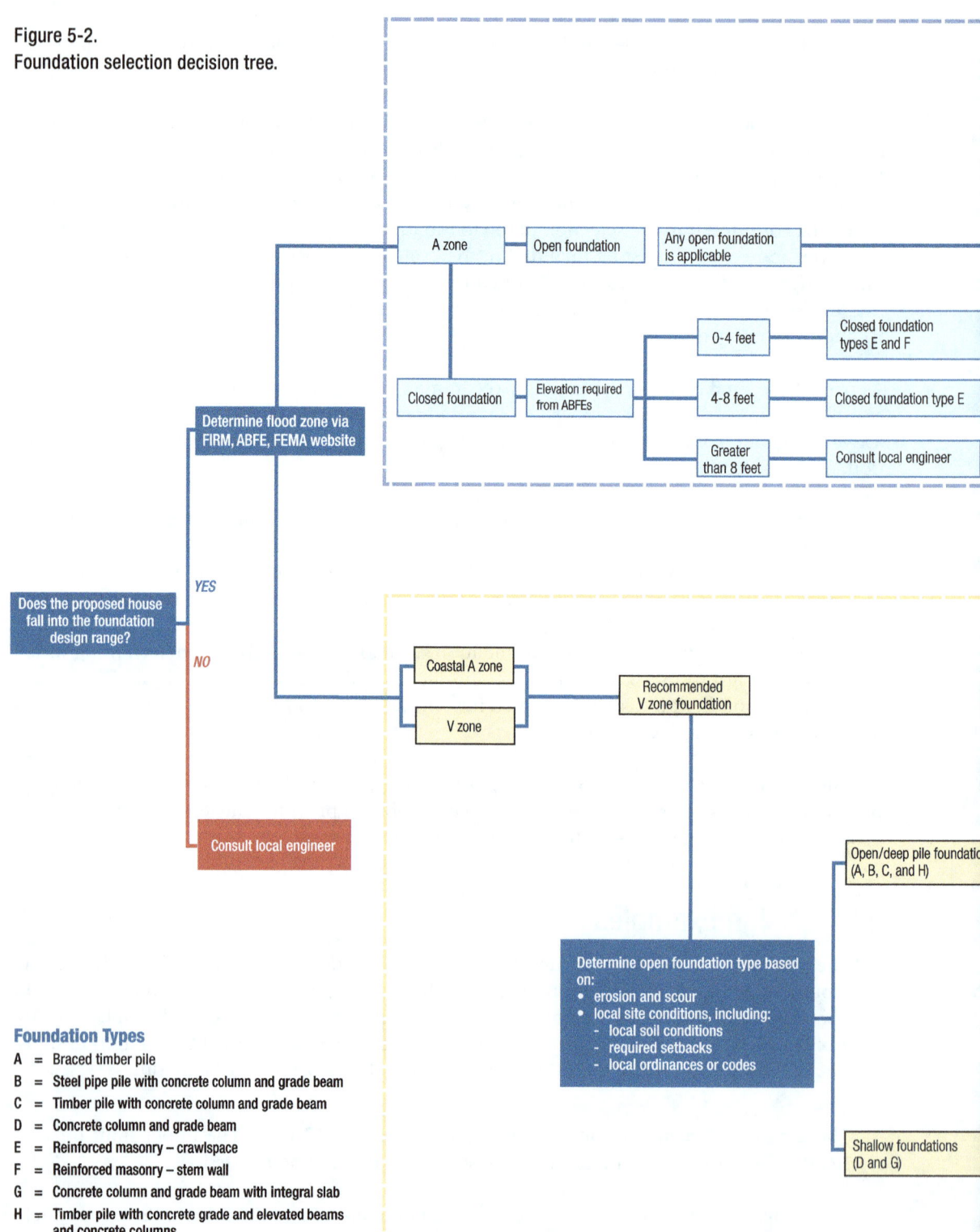

Foundation Types

A = Braced timber pile
B = Steel pipe pile with concrete column and grade beam
C = Timber pile with concrete column and grade beam
D = Concrete column and grade beam
E = Reinforced masonry – crawlspace
F = Reinforced masonry – stem wall
G = Concrete column and grade beam with integral slab
H = Timber pile with concrete grade and elevated beams and concrete columns

FOUNDATION SELECTION

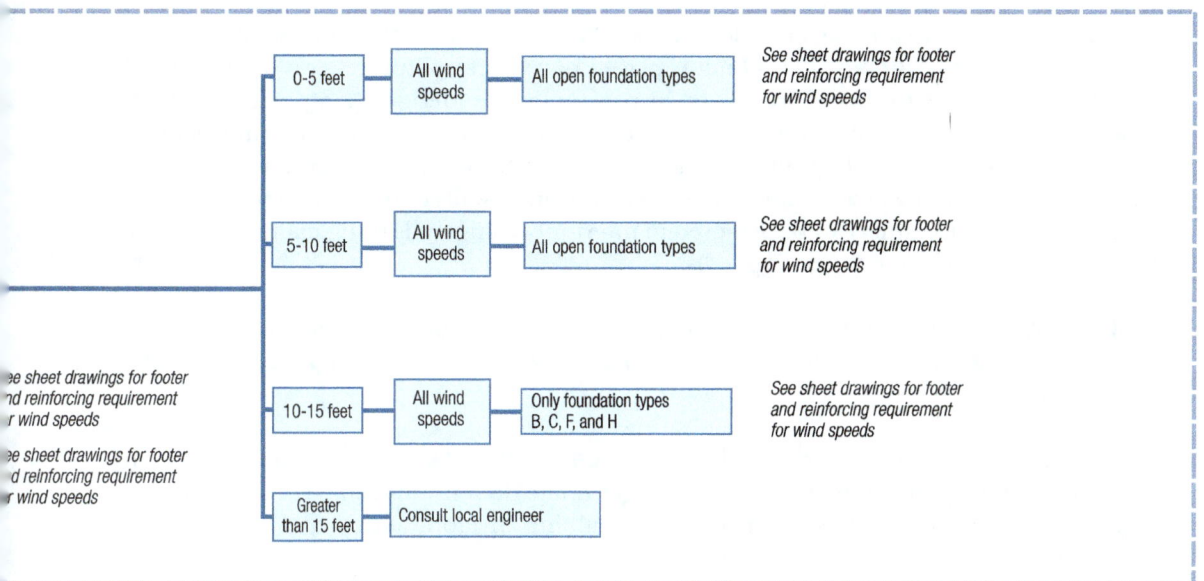

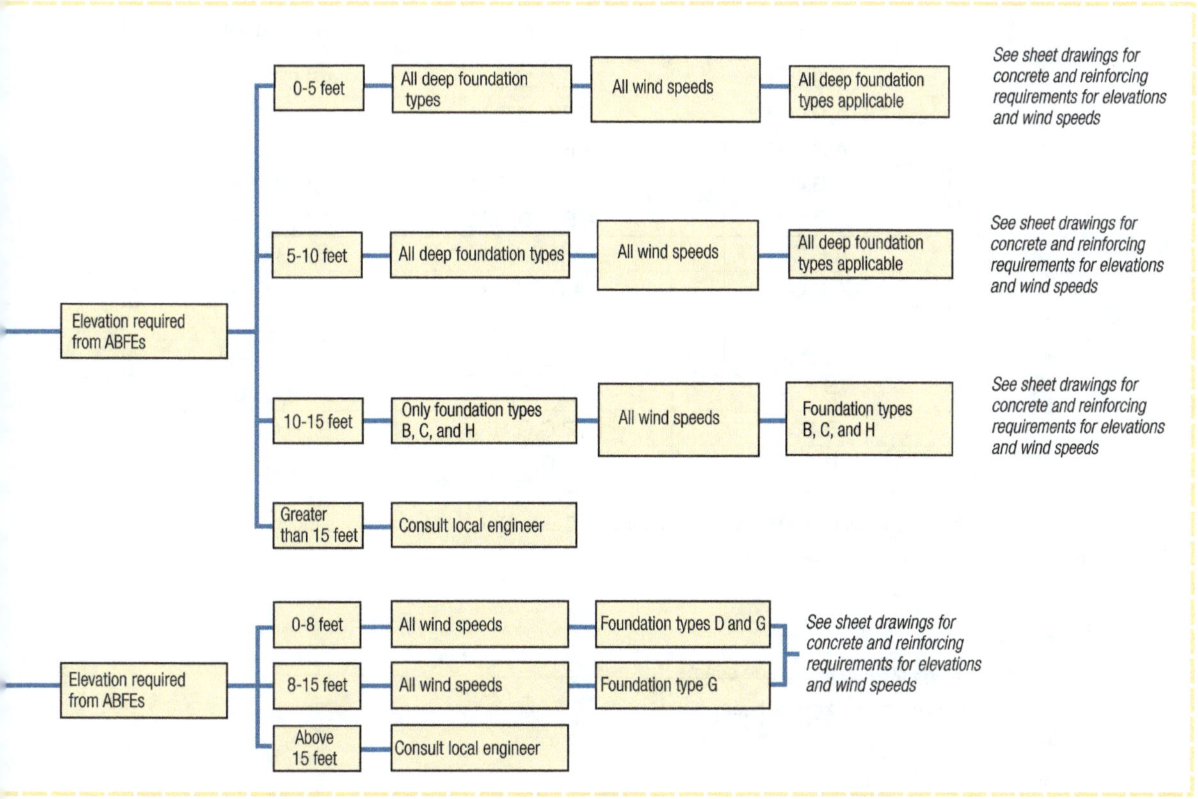

5 FOUNDATION SELECTION

3. **The "tributary load depth" of the roof framing shall not exceed 23 feet, including the 2-foot maximum roof overhang.** This limit is placed to restrict uplift forces on the windward foundation elements to those forces used in developing the design. As a practical matter, clear span roof trusses are rarely used on roofs over 42 feet deep; therefore, this limit should not be unduly restrictive. The roof framing that consists of multiple spans will require vertical load path continuity down through the interior bearing walls to resist uplift forces on the roof. Load path continuity can be achieved in interior bearing walls using many of the same techniques used on exterior bearing walls.

4. **On the perimeter foundation wall designs (Cases E and F), foundation shear walls must run the full depth of the building module, and shear walls can not be spaced more than 42 feet apart.**

5. **All foundation modules shall be at least 24 feet deep and at least 24 feet long.** Although the basic module is limited to 42 feet long, longer home dimensions can be developed, provided that the roof does not extend beyond the building envelope as depicted in Figure 2 of the Introduction.

Table 5-1a. Foundation Design Cases for One-Story Homes Based on Height of Elevation and Wind Velocity

	Height (H) (ft)	Wind Velocity of 120 to 150 (mph)		
		V Zone	Coastal A Zone*	Non-Coastal A Zone
One-Story Dwelling	< 4	A,B,C,H	A,B,C,D,G,H	A,B,C,D,E,F,G,H
	5	A,B,C,H	A,B,C,D,G,H	A,B,C,D,E,G,H
	6	A,B,C,H	A,B,C,D,G,H	A,B,C,D,E,G,H
	7	A,B,C,H	A,B,C,D,G,H	A,B,C,D,E,G,H
	8	A,B,C,H	A,B,C,D,G,H	A,B,C,D,E,G,H
	9	A,B,C,H	A,B,C,G,H	A,B,C,G,H
	10	A,B,C,H	A,B,C,G,H	A,B,C,G,H
	11	B,C,H	B,C,G,H	B,C,G,H
	12	B,C,H	B,C,G,H	B,C,G,H
	13	B,C,H	B,C,G,H	B,C,G,H
	14	B,C,H	B,C,G,H	B,C,G,H
	15	B,C,H	B,C,G,H	B,C,G,H

* In the Coastal A zone, the tops of all footings and grade beams in Cases D and G foundations must be placed below the maximum estimated erosion and scour depth.

Foundation Types

A = Braced timber pile

B = Steel pipe pile with concrete column and grade beam

C = Timber pile with concrete column and grade beam

D = Concrete column and grade beam

E = Reinforced masonry – crawlspace

F = Reinforced masonry – stem wall

G = Concrete column and grade beam with integral slab

H = Timber pile with concrete grade and elevated beams and concrete columns

Table 5-1b. Foundation Design Cases for Two-Story Homes Based on Height of Elevation and Wind Velocity

	Height (H) (ft)	Wind Velocity of 120 to 150 (mph)		
		V Zone	Coastal A Zone*	Non-Coastal A Zone
Two-Story Dwelling	< 4	A,B,C,H	A,B,C,D,G,H	A,B,C,D,E,F,G,H
	5	A,B,C,H	A,B,C,D,G,H	A,B,C,D,E,G,H
	6	A,B,C,H	A,B,C,D,G,H	A,B,C,D,E,G,H
	7	A,B,C,H	A,B,C,D,G,H	A,B,C,D,E,G,H
	8	A,B,C,H	A,B,C,D,G,H	A,B,C,D,E,G,H
	9	A,B,C,H	A,B,C,G,H	A,B,C,G,H
	10	A,B,C,H	A,B,C,G,H	A,B,C,G,H
	11	B,C,H	B,C,G,H	B,C,G,H
	12	B,C,H	B,C,G,H	B,C,G,H
	13**	B,C,H	B,C,H	B,C,H
	14**	B,C,H	B,C,H	B,C,H
	15**	B,C,H	B,C,H	B,C,H

* In the Coastal A zone, the tops of all footings and grade beams in Cases D and G foundations must be placed below the maximum estimated erosion and scour depth.

** Some foundation designs are not appropriate for two-story homes for a design wind speed of 150 mph. See individual design drawings for more details.

Foundation Types

A = Braced timber pile

B = Steel pipe pile with concrete column and grade beam

C = Timber pile with concrete column and grade beam

D = Concrete column and grade beam

E = Reinforced masonry – crawlspace

F = Reinforced masonry – stem wall

G = Concrete column and grade beam with integral slab

H = Timber pile with concrete grade and elevated beams and concrete columns

Figure 5-3.
"T" shaped modular design.
Note A: Overall building dimensions can exceed 42 feet. The vertical dimensions from the eave to the ridge roof shall not exceed 23 feet.

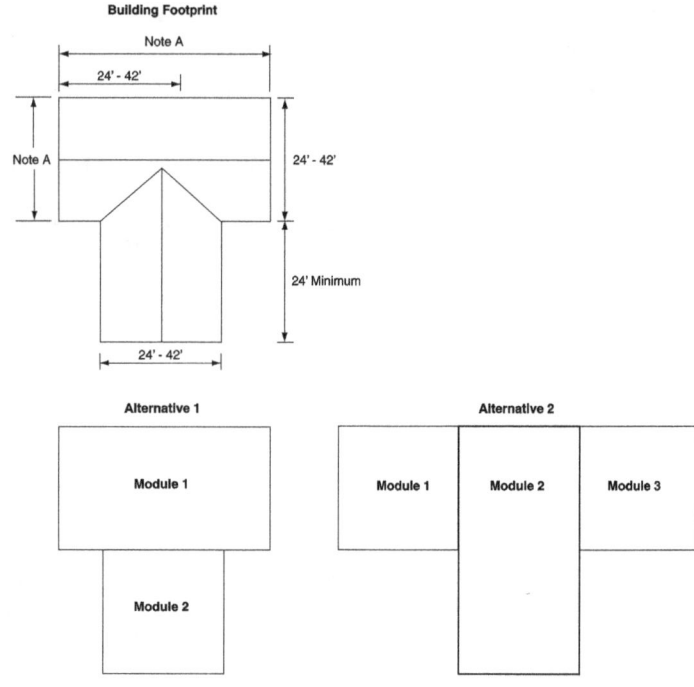

Figure 5-4.
"L" shaped modular design.
Note A: Overall building dimensions can exceed 42 feet. The vertical dimensions from the eave to the ridge roof shall not exceed 23 feet.

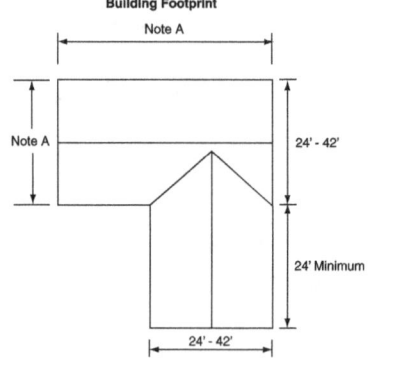

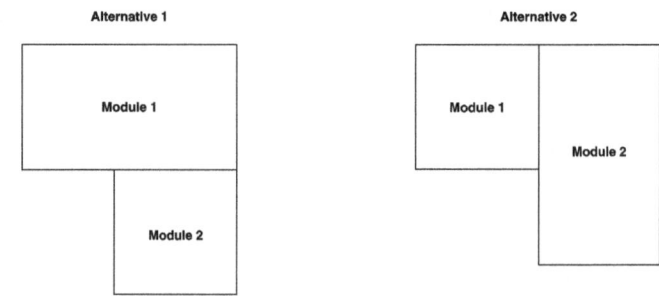

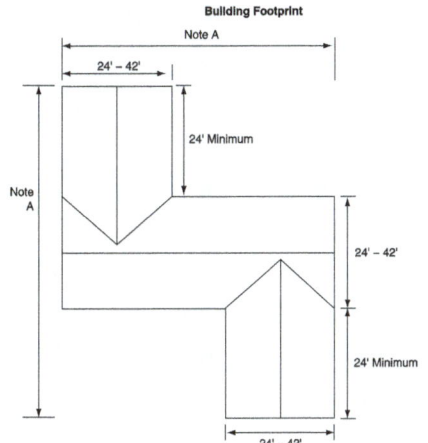
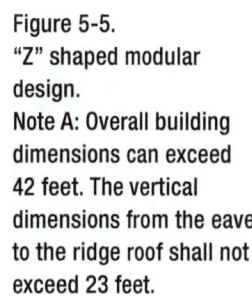

Figure 5-5.
"Z" shaped modular design.
Note A: Overall building dimensions can exceed 42 feet. The vertical dimensions from the eave to the ridge roof shall not exceed 23 feet.

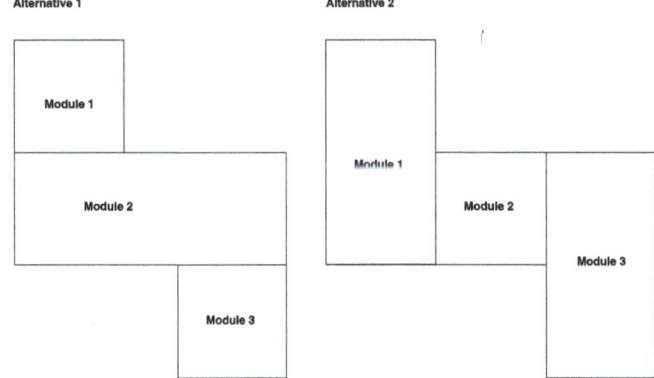

RECOMMENDED RESIDENTIAL CONSTRUCTION
FOR COASTAL AREAS

Building on Strong and Safe Foundations

A. Foundation Designs

This appendix contains the Case A through Case G drawings originally developed for the July 2006 edition of FEMA 550 and the new Case H drawings developed for this Second Edition of FEMA 550.

Case H drawings did not appear in the First Edition of FEMA 550. The drawings developed for the Case H designs vary in style and format from the original FEMA 550 designs for Cases A through G. Because of the differences in style and format, and the fact that thousands of copies of the July 2006 edition of FEMA 550 have been distributed, the Case H drawings are presented as standalone drawings. Separate title sheets and general notes for the Case H drawings are presented that are not intended to be used with the original drawings for Cases A through G.

Drawing
Title Sheet & Drawing Index
General Notes
Wood Beam Connections to Concrete Columns
Case A - Open Foundation Driven Treated Timber Pile
Case B - Open Foundation Open Ended Steel Pile
Case C - Open Foundation Driven Timber Pile & Concrete Pile
Case D - Open Foundation Continuous Footing & Concrete Column
Case G - Open Foundation Coastal Grade Beam/ Waffle
Case E - Closed Foundation Crawl Space Foundation
Case F - Closed Foundation Backfilled Stemwall Foundation

 Drawing Index

Drawing Number	Sheet Number	Revision Number
T-1	1 of 31	1
GN-1	2 of 31	1
GN-2 to GN-6	3 to 7 of 31	1
A-1 to A-3	8 to 10 of 31	1
B-1 to B-5	11 to 15 of 31	1
C-1 to C-6	16 to 21 of 31	1
D-1 to D-3	22 to 24 of 31	1
G-1, G-2	25 to 26 of 31	1
E-1 to E-3	27 to 29 of 31	1
F-1, F-2	30 to 31 of 31	1

Foundation Designs
Title Sheet & Drawing Index

DRAWING NO.: T-1	SHEET 1 of 31
DATE: August 8, 2006	
REVISED:	REV. 9

1 General

The use of the foundation designs of this manual are not intended to convey any particular sequence or procedure. The respective builder and/or contractor shall be responsible for taking adequate means and measures to insure the stability of the building and it's components during construction. These shall include, but are not limited to; necessary shoring, sheeting, temporary bracing, dewatering, etc. This manual has been provided to assist in the reconstruction efforts after Hurricane Katrina. Builders, architects, or engineers using this manual assume responsibility for the resulting designs.

2 Foundation

Foundation designs are based on minimum values believed to exist in most areas of the Gulf Coast Region. Actual site soil characteristics shall be verified as being in compliance with the assumed values stated and in compliance with local building codes. Parties using the foundation designs may wish to retain the services of a Geotechnical Engineer to verify local conditions and to provide site specific values.

Soil supported foundations are based on a presumptive allowable soil pressure of 1500 pounds per square foot. In non V zones, compacted structural fill may be used. It is recommended that structural fill be placed in maximum 6" layers and compacted to 95% density as measured by the Modified Proctor method. It is also recommended, and may be required by local code, that field compaction tests be performed by a testing agent. Pile supported foundations are based on presumptive values indicated on the detail drawings.

It is advised that the location of all underground utilities be verified prior to commencing any foundation work. Proper care should be taken during any excavation work as uncharted underground utilities may exist in the area.

3 Concrete

All structural concrete, for foundations, slabs on grade, columns and beams, etc., shall be a plant batched ready-mix. The concrete mix shall be of standard weight aggregates able to achieve a 28 day compressive strength of 4000 psi. The use of Calcium Chlorides shall not be permitted. Use of water reducing agents / superplasticizer are recommended. Use of plastic fiber additives for exposed concrete slabs at grade is also recommended.

All concrete work shall comply with the requirements of ASTM Standard C94 for the measuring, mixing, transporting, placing, etc. Concrete tickets should be time stamped when the concrete is batched. The maximum time allowed, from when the mixing water is added to the time the concrete is deposited should not exceed one and one half hours. Use of concrete that is not in compliance with the above can result in significant reduction in concrete strength and performance. Concrete shall be placed with due regard to extreme temperature conditions. Refer to ACI 305 " Hot Weather Concreting" and ACI 306 "Cold Weather Concreting" for guidance in placing concrete during weather extremes.

Refer to ACI 117 and ACI 347R for guidance in planning Form work design and for finish standards and requirements.

A standard hook shall be provided at the termination of all beam top reinforcing bars. Corner bars with full tension laps, matching the size and spacing of the continuous bar shall be provided.

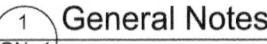 General Notes

Reinforcing steel shall be of ASTM A615 Grade 60 deformed bars, free from oil, scale and excessive rust. Reinforcing shall be detailed, fabricated and placed in accordance with the typical bending diagrams, placing requirements and details of the latest editions of the American Concrete Institute (ACI) standards and specifications. These publications include ACI 318, ACI 301, ACI 304, ACI 315 and ACI SP-68. Minimum Cover for reinforcing is; 3" for concrete cast against soil and 2" for concrete that is exposed to weather. It is recommended that the owner utilize a concrete testing agent to verify concrete strength. Reinforcing shall be secured in place by the use of metal ties and supported by metal bolsters, chairs, spacers and other devices.
Welded wire fabric shall conform to ASTM A185 and be furnished in flat sheets only.
Standard Lap lengths are as follows:

Bar size	Lap length	Bar size	Lap length
#3	18"	#7	42"
#4	24"	#8	48"
#5	30"	#9	54"
#6	36"	#10	60"

Reinforcement Lap Splice Table

4 Reinforced Masonry Construction

All concrete masonry unit (CMU) construction shall conform to the recommendations of the "BUILDING CODE REQUIREMENTS FOR MASONRY STRUCTURES", ACI 530-02/ASCE 5-02/TMS 402-02 and ACI 530.1-02 Specifications for Masonry Structures.
All steel reinforcement shall conform to ASTM A615 Grade 60 material.
All steel reinforcement lap lengths shall conform to table shown in SECTION 3.
The compressive strength of masonry, f'm, shall be a minimum of 1500 psi.
All reinforced cells shall be solidly filled with grout. All cells that have anchors, embedded plates, etc., shall be filled with grout. All cells within 8" above grade and all cells below grade shall be filled with grout. Grout shall have a 28-day compressive strength of 3000 psi. and be of pea gravel aggregates.
All CMU shall be laid in a running bond with a full mortar bed.

5 Structural Steel

Structural shapes, anchor bolts, etc. that are indicated to be galvanized, shall be hot dip galvanized after fabrication per ASTM A 368, A123 to G90 standard.

6 Connections

Connections to wood framing are based on lumber having a minimum specific gravity of 0.55. See sheet GN-2 to GN-6 for these details.

7 Driven Piles

Treated Timber Piles shall conform to latest revision of ASTM Specification D-25. Piles shall be of Southern Pine or Douglas Fir clean or rough peeled. Piles also shall have a minimum tip diameter of 8 inches and a minimum butt diameter of 12 inches as measured 3 feet from the end of the pile.

All treated timber piles shall be preservative treated. Preservative retention shall be 0.61 pounds per cubic foot. Pile preservative treatment shall comply with AWPA C3 standards.

Treat cut ends and drilled holes of treated timber piles with the same penetrating preservative as base treatment chemical. All piles shall be uniformly spaced along width and depth of building. Variations of +/- 1' are allowed to address site conditions, provided piles are spaces a minimum of 4 pile diameters unless otherwise noted.

General Notes

DRAWING NO.: GN-1 | SHEET 2 OF 31
DATE: August 8, 2006
REVISED: | REV. 9

FEMA

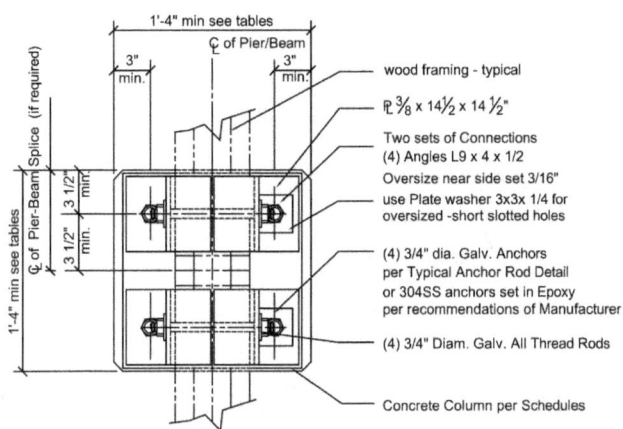

Captive-Clamped Beam (CCB) - Typical Connection Plan Detail
1/GN-2 SCALE: NTS

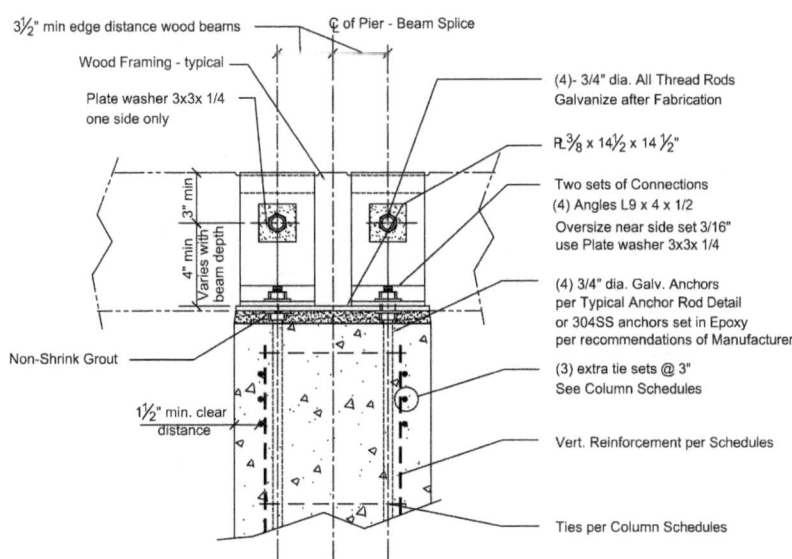

Captive-Clamped Beam (CCB) - Typical Connection Elevation
3/GN-2 SCALE: NTS

Note: All steel material min. ASTM grade A36 and Galvanized after fabrication u.n.o. Connection design is based on wood species having a specific gravity of at least 55% such as southern yellow pine.

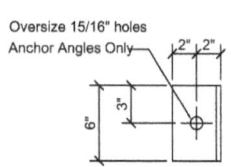
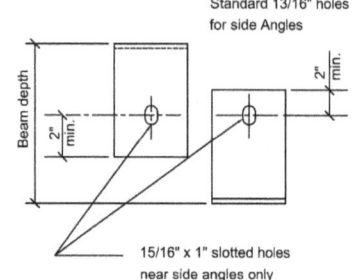

Captive-Clamped Beam (CCB) - Typical Connection Angle Detail
(2 / GN-2) SCALE: NTS

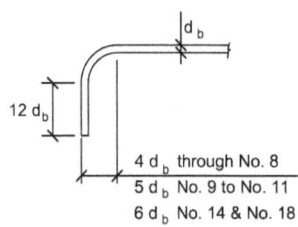
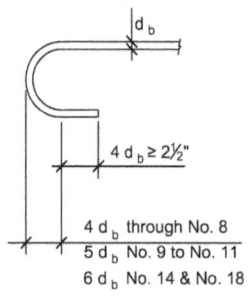

Standard Hooks
(4 / GN-2) SCALE: NTS

Connection Details

DRAWING NO.: GN-2 | SHEET 3 OF 31
DATE: August 8, 2006
REVISED | REV. 9

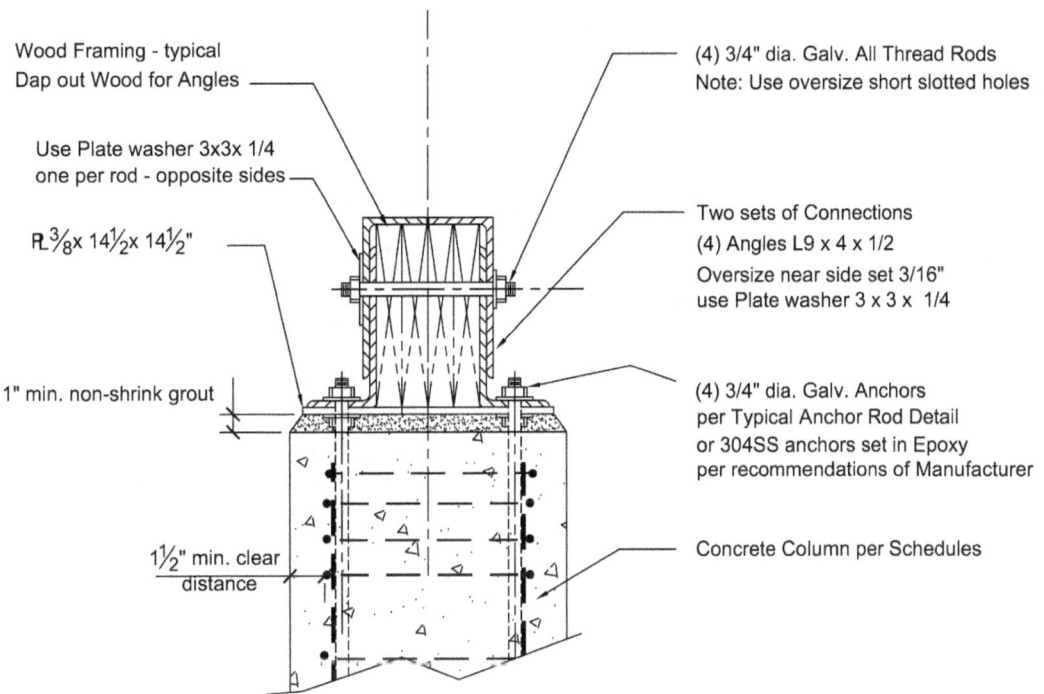

Captive-Clamped Beam (CCB) - Typical Connection Section
GN-3 / 1   SCALE: NTS

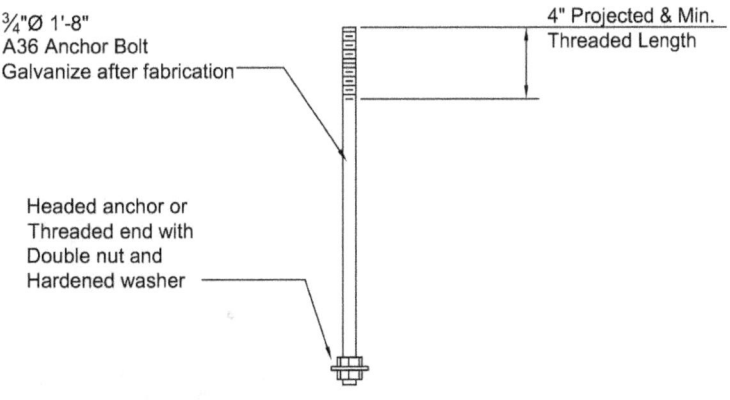

Typical Anchor Rod Detail - Cast-In Place
GN-3 / 3   SCALE: NTS

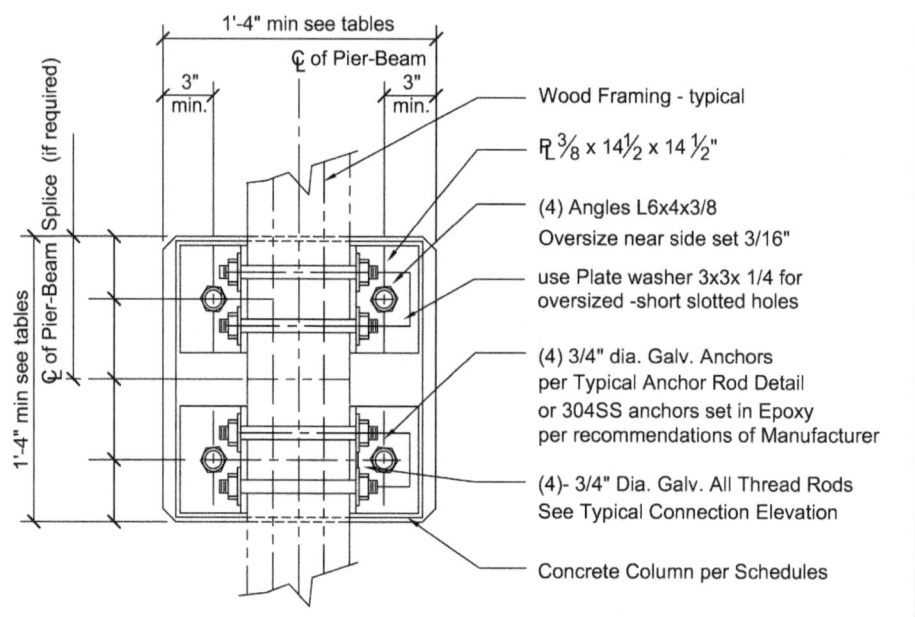

2/GN-3 Double Angle Beam (DAB) - Typical Connection Plan Detail
SCALE: NTS

Connection Details	
DRAWING NO.: GN-3	SHEET 4 OF 31
DATE: August 8, 2006	
REVISED:	REV. 9

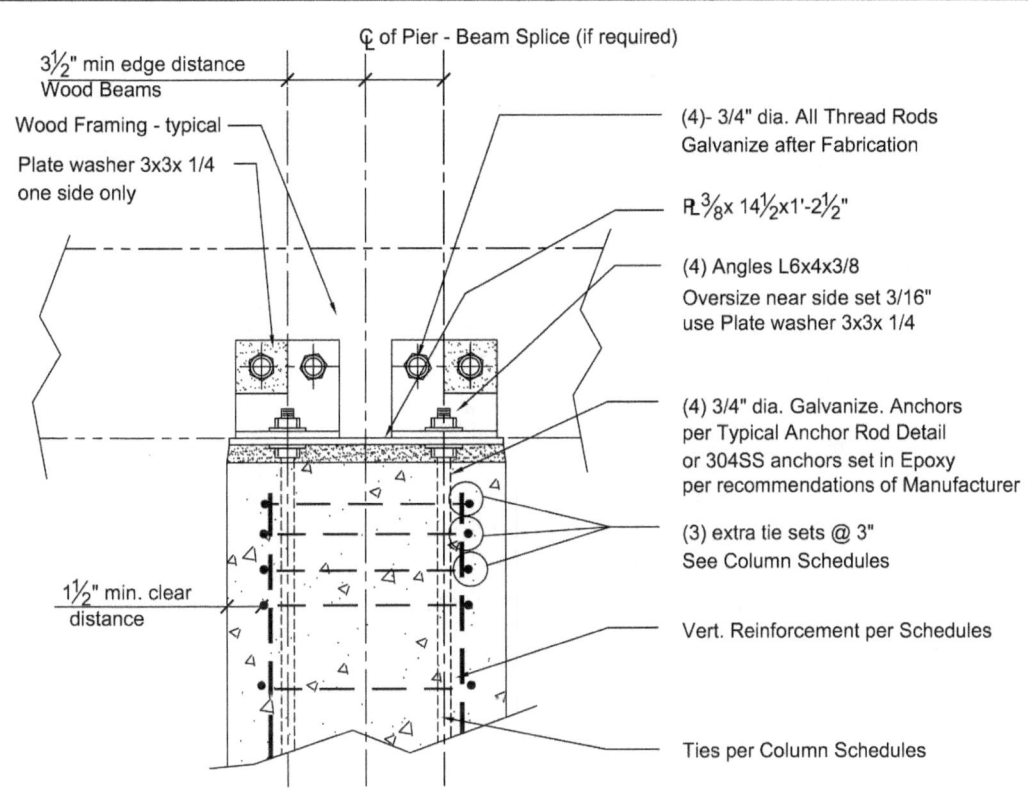

Double Angle Beam (DAB) - Typical Connection Section
1/GN-4 SCALE: NTS

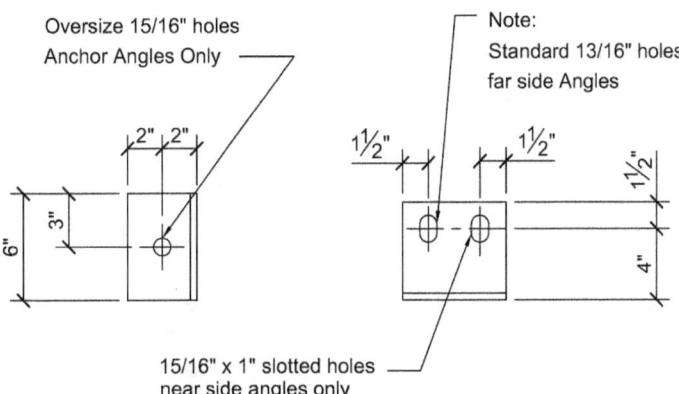

Double Angle Beam (DAB) - Typical Connection Section
3/GN-4 SCALE: NTS

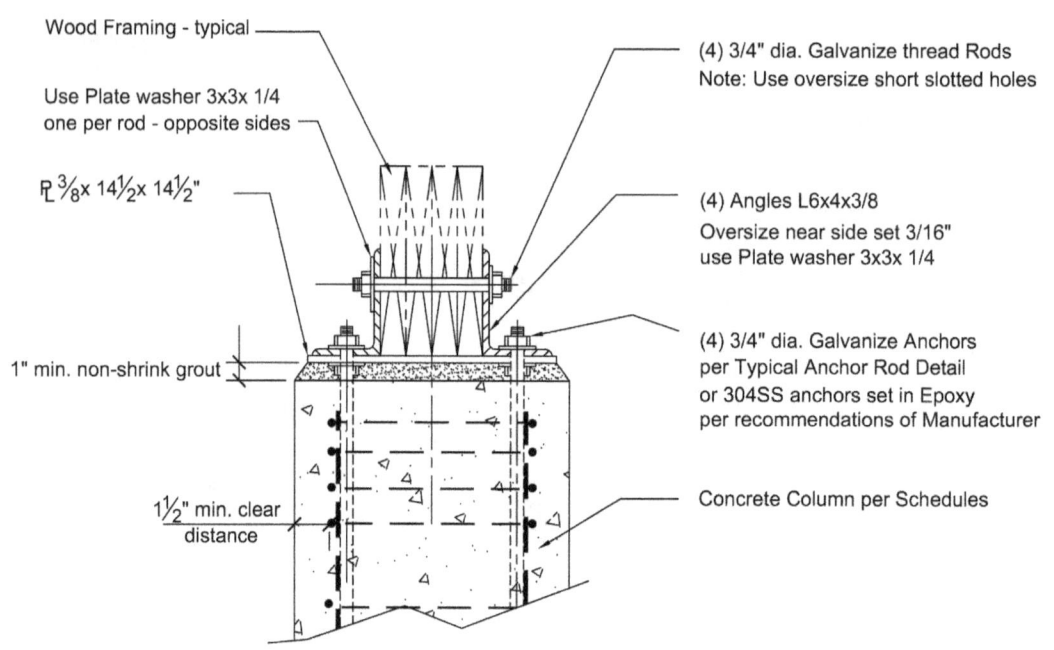

2/GN-4 Double Angle Beam (DAB) - Typical Connection Section
SCALE: NTS

Connection Details	
DRAWING NO.: GN-4	SHEET 5 OF 31
DATE: August 8, 2006	
REVISED:	REV. 9

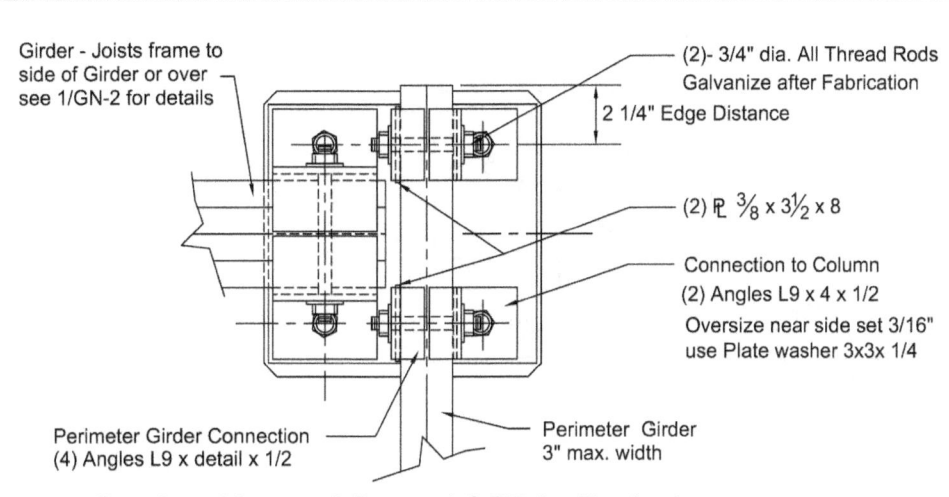

Captive-Clamped Beam (CCB) - Typical Corner Connection Plan Detail
1/GN-5 SCALE: NTS

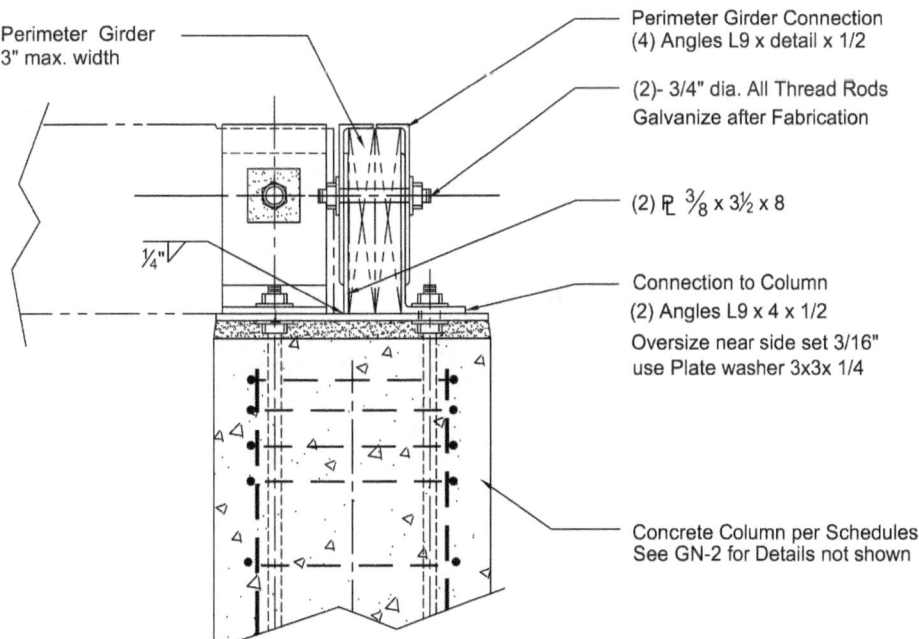

Captive-Clamped Beam (CCB) - Typical Corner Connection Section
2/GN-5 SCALE: NTS

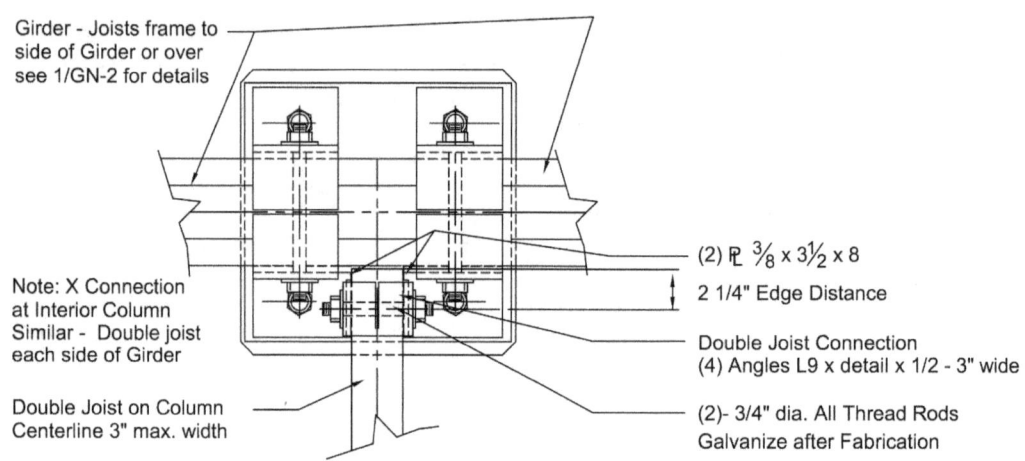

Captive-Clamped Beam (CCB) - Typical T Connection Plan Detail
3/GN-5 SCALE: NTS

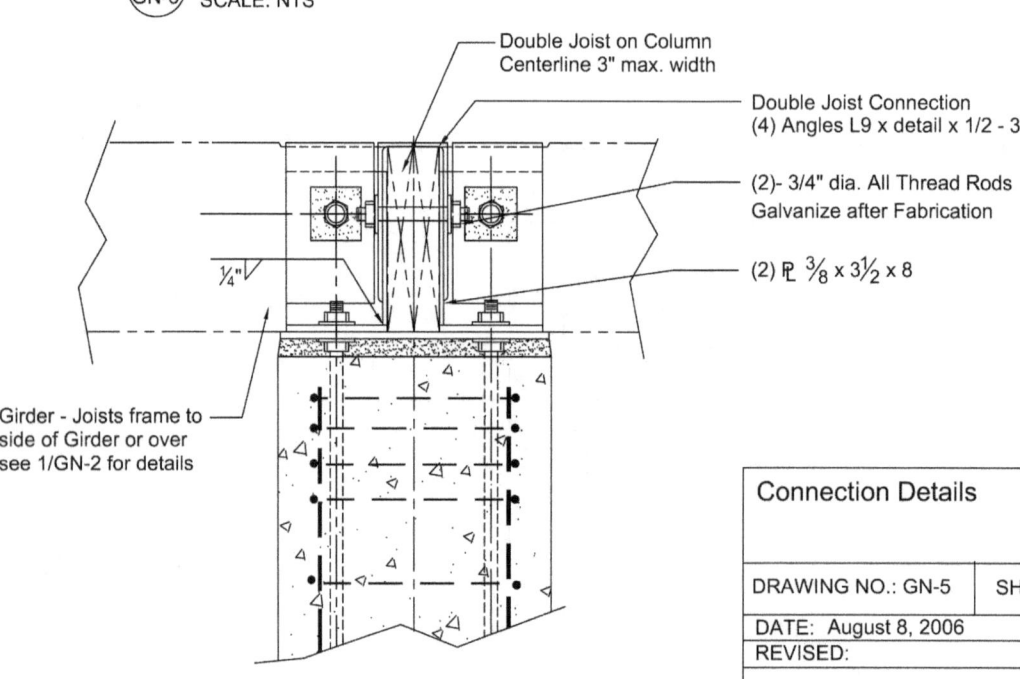

Captive-Clamped Beam (CCB) - Typical "T" Connection Section
4/GN-5 SCALE: NTS

Connection Details

DRAWING NO.: GN-5	SHEET 6 OF 31
DATE: August 8, 2006	
REVISED:	REV. 9

FEMA

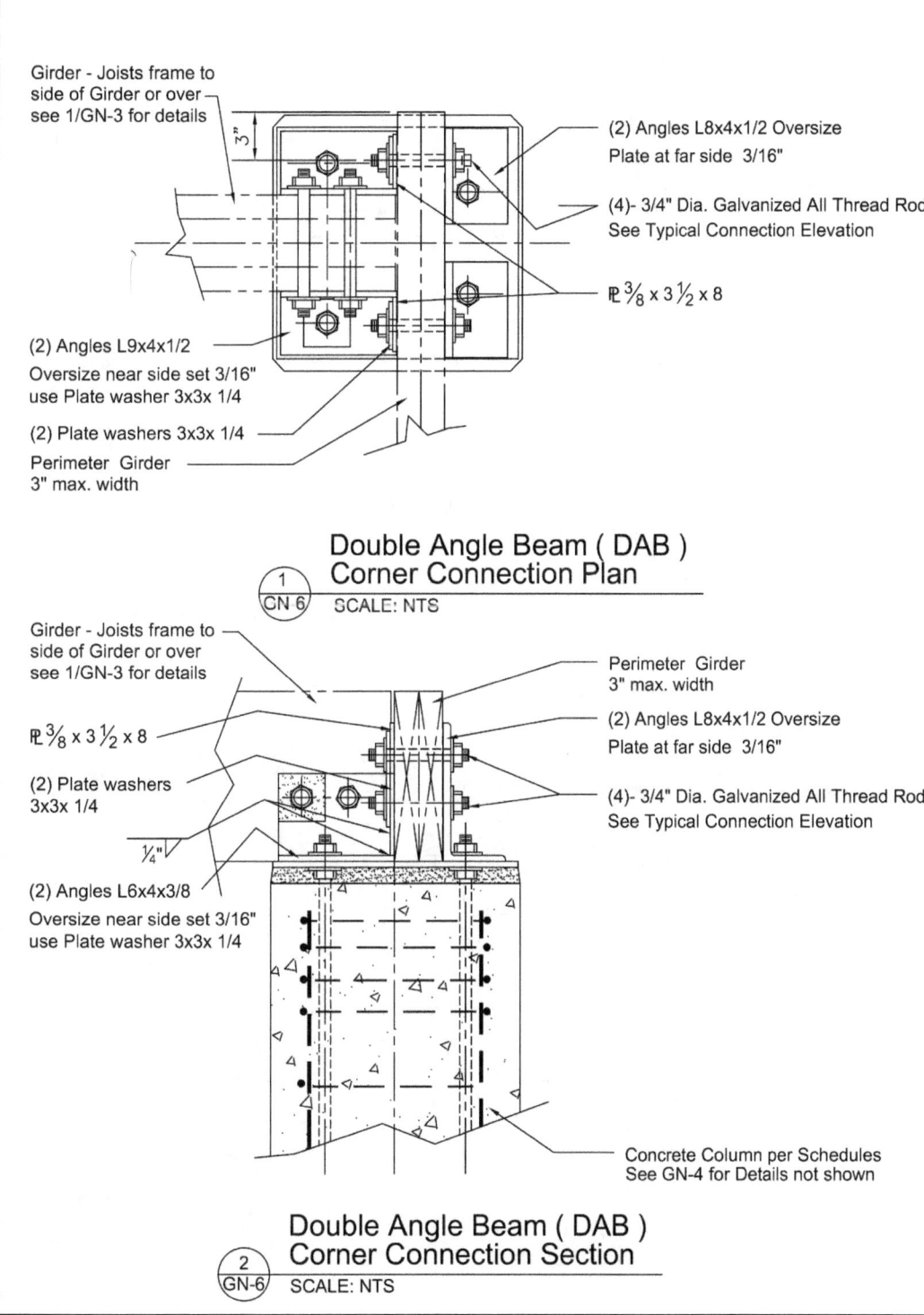

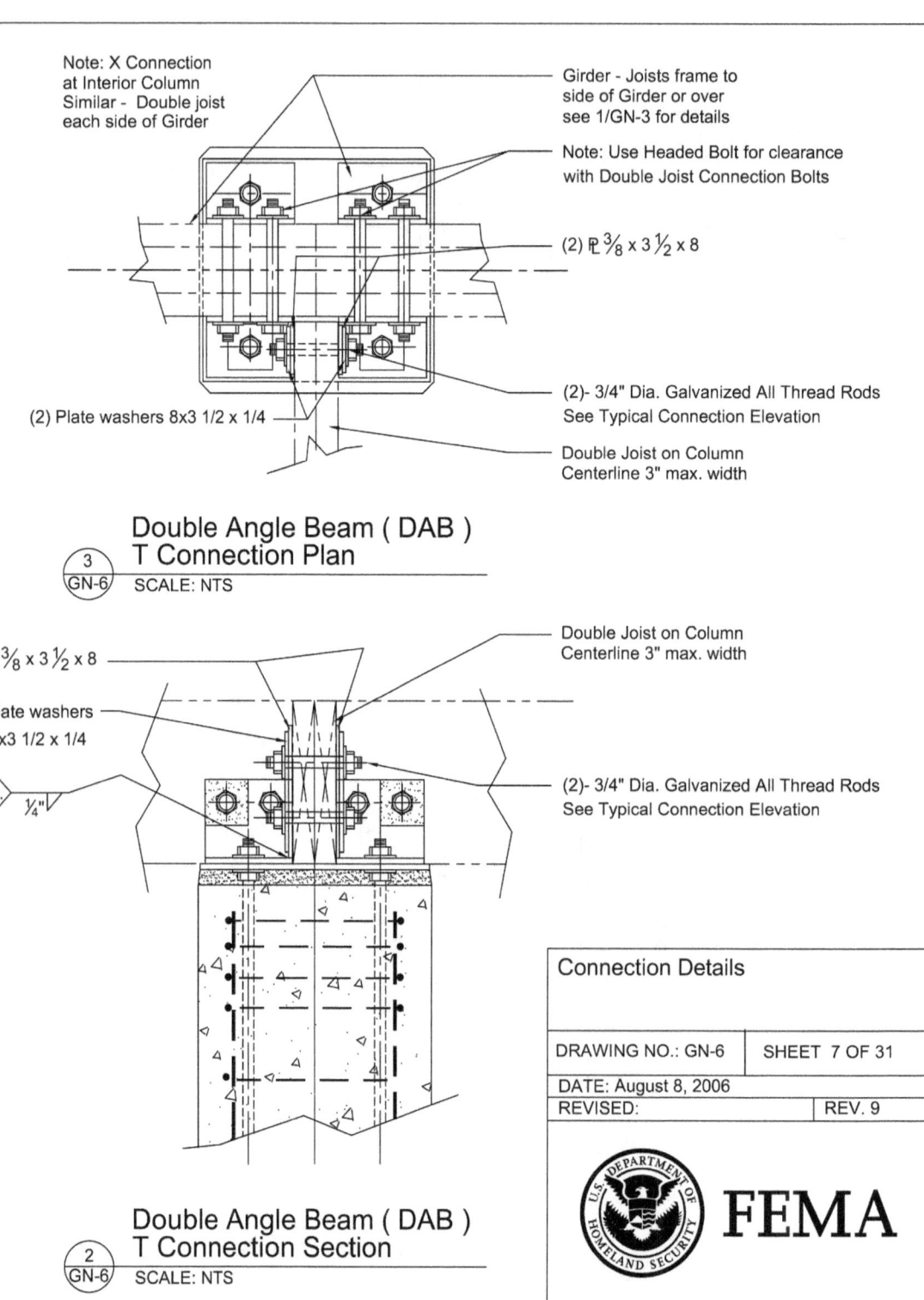

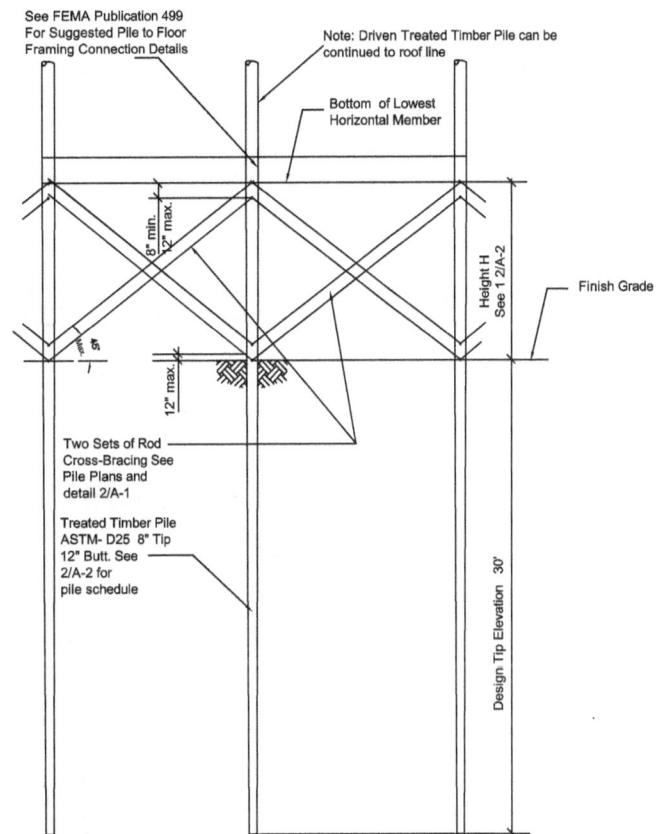

Case A Open Foundation - Driven Treated Timber Pile

1/A-1 **Pile Profile - Double Rod Cross Bracing Shown**
SCALE: NTS

Piles are assumed to have the following capacities:
Compression   14000 pounds
Tension   9300 pounds
Lateral   4000 pounds

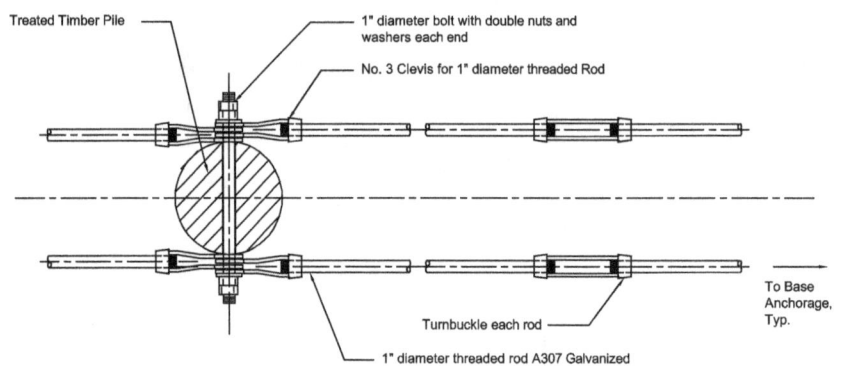

Case A Open Foundation - Driven Treated Timber Pile
Rod Bracing Detail
(2/A-1) SCALE: NTS

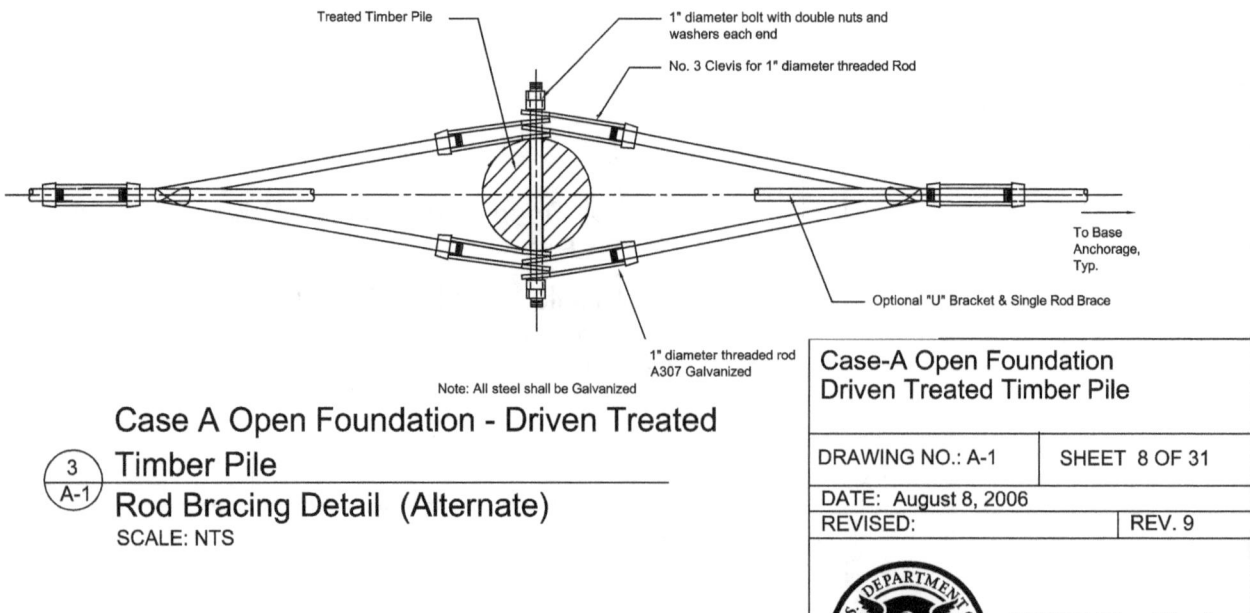

Case A Open Foundation - Driven Treated Timber Pile
Rod Bracing Detail (Alternate)
(3/A-1) SCALE: NTS

Note: All steel shall be Galvanized

Case-A Open Foundation Driven Treated Timber Pile	
DRAWING NO.: A-1	SHEET 8 OF 31
DATE: August 8, 2006	
REVISED:	REV. 9

U.S. DEPARTMENT OF HOMELAND SECURITY — FEMA

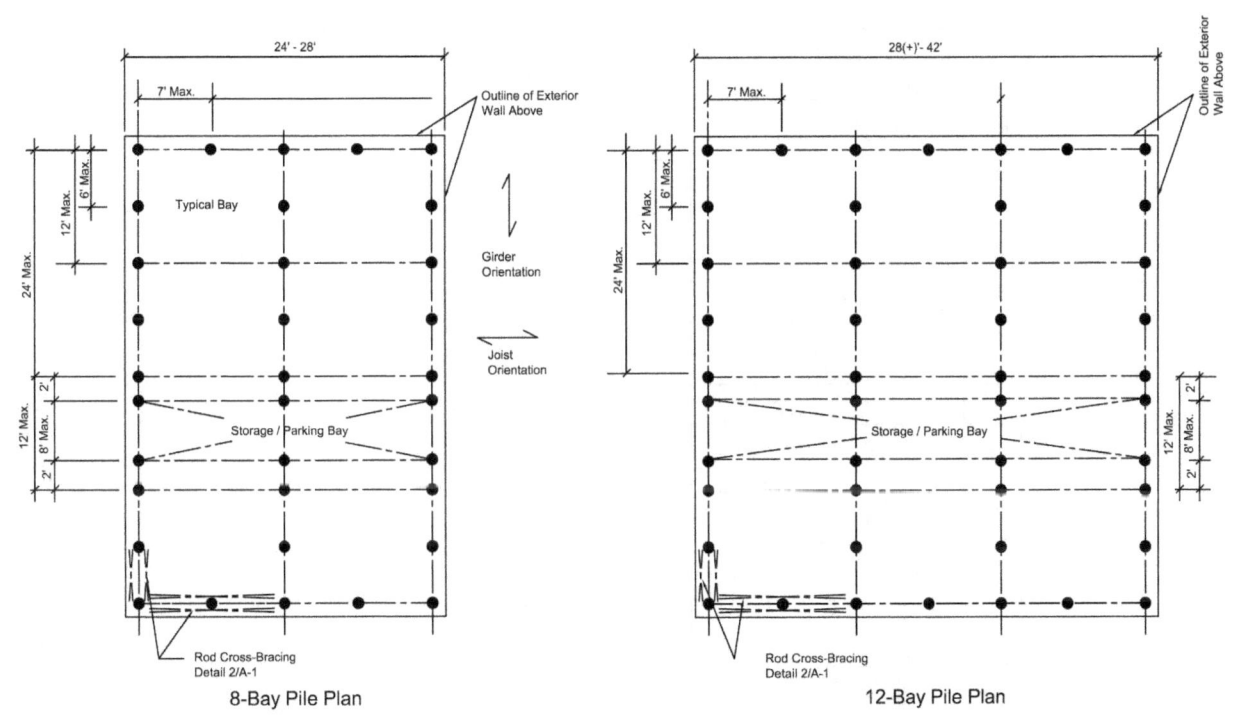

Case A Open Foundation - Driven Treated Timber Pile
Pile Plan - 1
SCALE: NTS
(1/A-2)

Girder Orientation ↑

Joist Orientation →

Braced Timber Pile Foundation				
Pile Schedule - One Story Structures				
"H"	Wind Speed			
	150 mph	140 mph	130 mph	120 mph
< 6 ft	Pile Plan 1	Pile Plan 1	Pile Plan 1	Pile Plan 1
8 ft	Pile Plan 1	Pile Plan 1	Pile Plan 1	Pile Plan 1
10 ft	Pile Plan 1	Pile Plan 1	Pile Plan 1	Pile Plan 1

Braced Timber Pile Foundation				
Pile Schedule - Two Story Structures				
"H"	Wind Speed			
	150 mph	140 mph	130 mph	120 mph
< 6 ft	Pile Plan 1	Pile Plan 1	Pile Plan 1	Pile Plan 1
8 ft	Pile Plan 1	Pile Plan 1	Pile Plan 1	Pile Plan 1
10 ft	Pile Plan 2	Pile Plan 2	Pile Plan 2	Pile Plan 2

Note:
1) Pile Bracing shall not be installed in areas where pile diameter is less than 8 inches
2) See Drawing A-3 for bracing schedule.

(2/A-2) **Case A - Table 1 Treated Timber Pile Pile Plan Schedule**

Case-A Open Foundation
Driven Treated Timber Pile

DRAWING NO.: A-2   SHEET 9 of 31
DATE: August 8, 2006
REVISED: REV. 8

FEMA

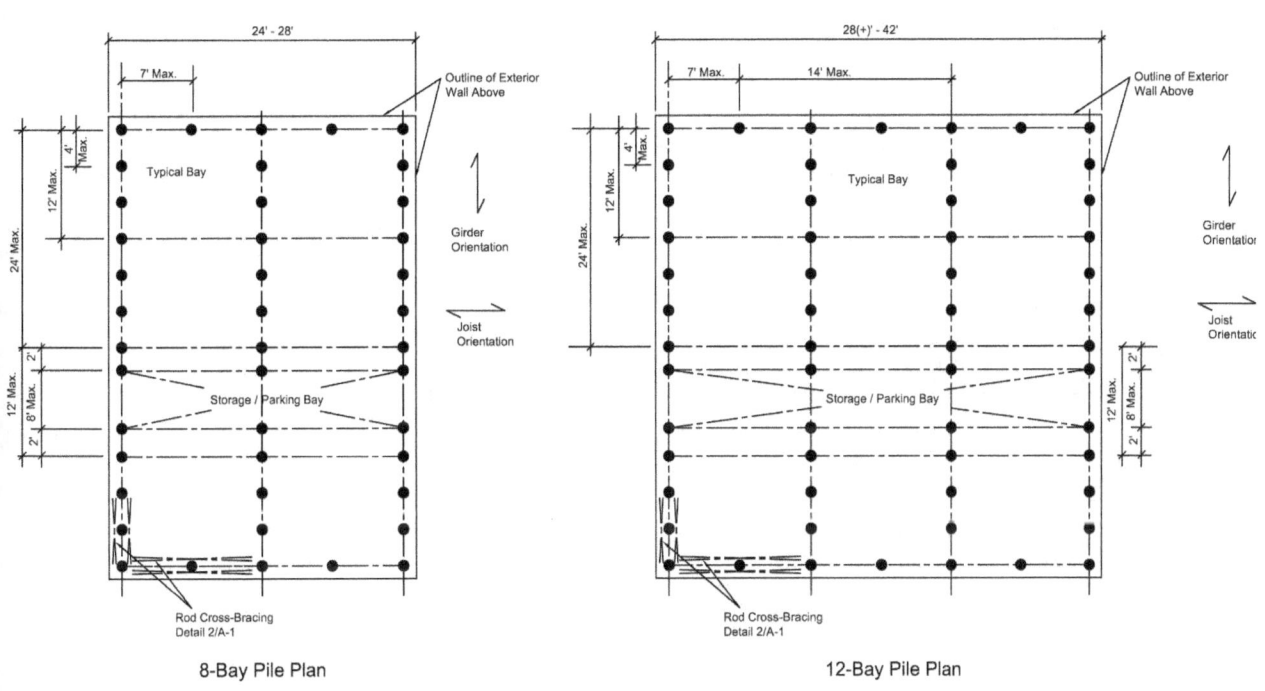

Case A Open Foundation - Driven Treated Timber Pile
Pile Plan - 2
SCALE: NTS

Bracing Notes:
1) Bracing quantities shown in Table 1 are for "pairs" of 1 inch diameter braces required for each 24 feet of home. Each brace pair shall be installed to form an "X" as shown on Detail 1/A-1 and connected to the pilings with 1 inch through bolts.
2) Bracing quantities shown in Table 1 shall be installed in both directions (across width and length of home). All corner piles and piles adjacent to framed openings in the first floor shall be braced. Bracing in other areas shall be uniformly distributed. Bracing in only one direction can be placed in other areas of the home to create storage/parking areas provided the total number of braces specified is installed.
3) Bracing shall be scaled for homes with other dimensions by the factor $(N/24) \times L$ where N is the number of braces shwon in Table 1 and L is the dimension of the home perpendicular to the braces. For example: a 32' by 24' wide one story home located in a 130 mph wind speed zone and elevated 8' requires 20 braces perpendicular to the 32' dimension ($(32'/24') \times 12$ braces) and 12 braces perpendicular to the 24' dimension.
4) Rod shall be installed at a maximum angle of 45 degrees to the horizontal and shall be connected to the top of the pile within 12" of the bottom girder and within 12" of exterior grade.
5) Each rod in the brace pairs shall be connected to a single bolt in the piling to create a double shear connection.
6) Up to two braces can be connected to a single pile (in both directions). Brace connection points shall be offset vertically 8" (min) to 12" (max).
7) Provide a tensioning turnbuckle for each rod brace.
8) Two rods and turnbuckles in a brace pair may be substituted with a single rod, turnbuckle and two "U" brackets. U brackets must have an inside diameter of no more than 1 inch greater than pile diameter at the brace point, must be capable of resisting a 7,500# working load without yielding and must transfer loads to the piling without creating torsion forces.
9) Pile Bracing shall not be installed in areas where pile diameter is less than 8 inches
10) * No bracing is required for H < 4 ft.

Braced Timber Pile Foundation				
Bracing Schedule - One Story Structures				
"H"	Wind Speed			
	150 mph	140 mph	130 mph	120 mph
< 6 ft *	12	11	10	10
8 ft	14	12	12	11
10 ft	16	14	14	13

Braced Timber Pile Foundation				
Bracing Schedule - Two Story Structures				
"H"	Wind Speed			
	150 mph	140 mph	130 mph	120 mph
< 6 ft *	18	17	16	15
8 ft	20	19	18	17
10 ft	25	24	23	22

(2/A-3) **Case A - Table 1 Driven Treated Timber Pile**
Pile Bracing Schedule

Case-A Open Foundation
Driven Treated Timber Pile

DRAWING NO.: A-3	SHEET 10 OF 31
DATE: August 8, 2006	
REVISED:	REV. 9

Square Column Size & Reinforcement Schedule

One Story

"H"	150 mph Size (sq)	150 mph Reinforcing	140 mph Size (sq)	140 mph Reinforcing	130 mph Size (sq)	130 mph Reinforcing	120 mph Size (sq)	120 mph Reinforcing
8 ft	16"	A	16"	A	16"	A	16"	A
10 ft	16"	C	16"	C	16"	B	16"	B
12 ft	18"	D	18"	D	16"	D	16"	D
15 ft	20"	A	20"	A	18"	D	18"	C

Square Column Size & Reinforcement Schedule

Two Story

"H"	150 mph Size (sq)	150 mph Reinforcing	140 mph Size (sq)	140 mph Reinforcing	130 mph Size (sq)	130 mph Reinforcing	120 mph Size (sq)	120 mph Reinforcing
8 ft	16"	B	16"	B	16"	B	16"	A
10 ft	16"	D	16"	C	16"	C	16"	B
12 ft	18"	D	18"	C	18"	C	18"	B
15 ft	20"	A	20"	A	18"	D	18"	D

Notes:
1) See Table 2, 3/B3 sheet for Column size and reinforcement details

1/B-1 Case B Open Foundation - Steel Pile/Concrete Column/Grade Beam Table 1 Column Schedule

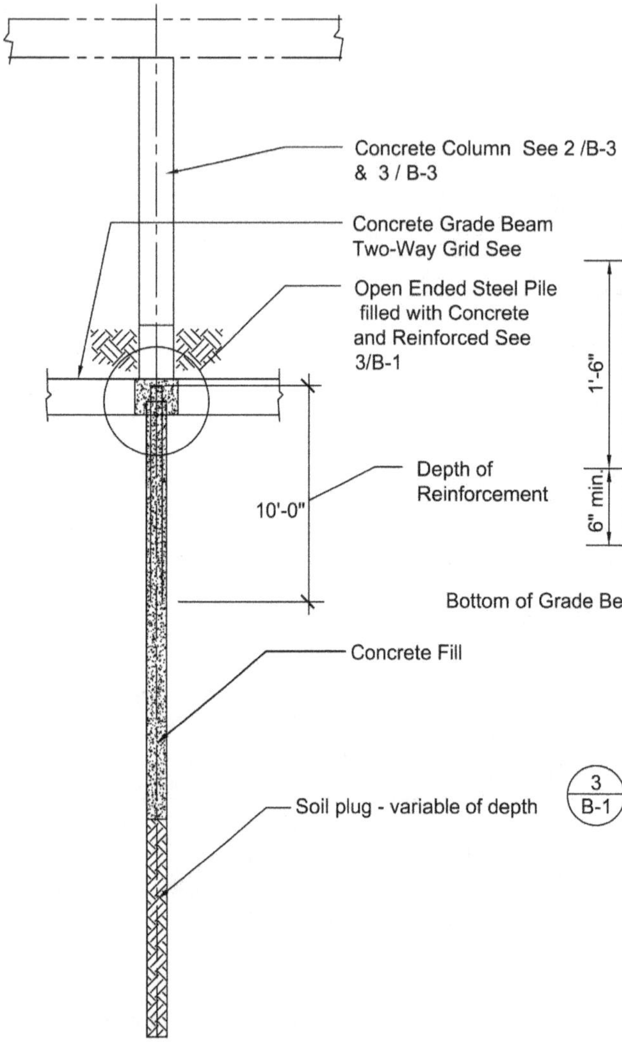

Pile Detail (2/B-1) — Case B Open Foundation - Steel Pile/Concrete Column/Grade Beam

Labels (left detail):
- Concrete Column See 2/B-3 & 3/B-3
- Concrete Grade Beam Two-Way Grid See
- Open Ended Steel Pile filled with Concrete and Reinforced See 3/B-1
- Depth of Reinforcement
- 10'-0"
- Bottom of Grade Beam
- Concrete Fill
- Soil plug - variable of depth

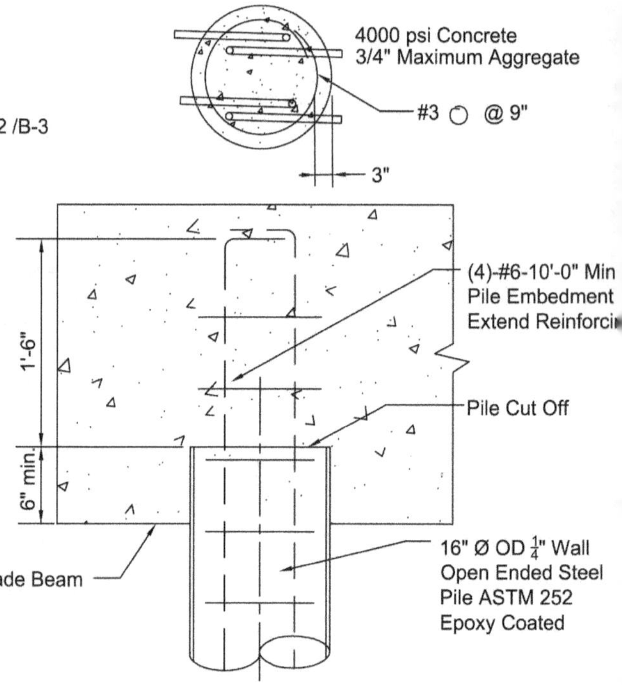

Pile Detail (3/B-1) — Case B Open Foundation - Steel Pile/Concrete Column/Grade Beam

Labels (right detail):
- 4000 psi Concrete 3/4" Maximum Aggregate
- #3 @ 9"
- 3"
- 1'-6"
- 6" min.
- (4)-#6-10'-0" Min Pile Embedment Extend Reinforcing
- Pile Cut Off
- 16" Ø OD 1/4" Wall Open Ended Steel Pile ASTM 252 Epoxy Coated

SCALE: NTS

Piles are assumed to have the following capacities:
- Compression 20,000 lbs.
- Tension 13,400 lbs.
- Lateral 8,000 lbs.

Case-B Open Foundation
Steel Pipe Pile with Concrete Column and Grade Beam

DRAWING NO.: B-1 | SHEET 11 OF 31
DATE: August 8, 2006
REVISED: | REV. 9

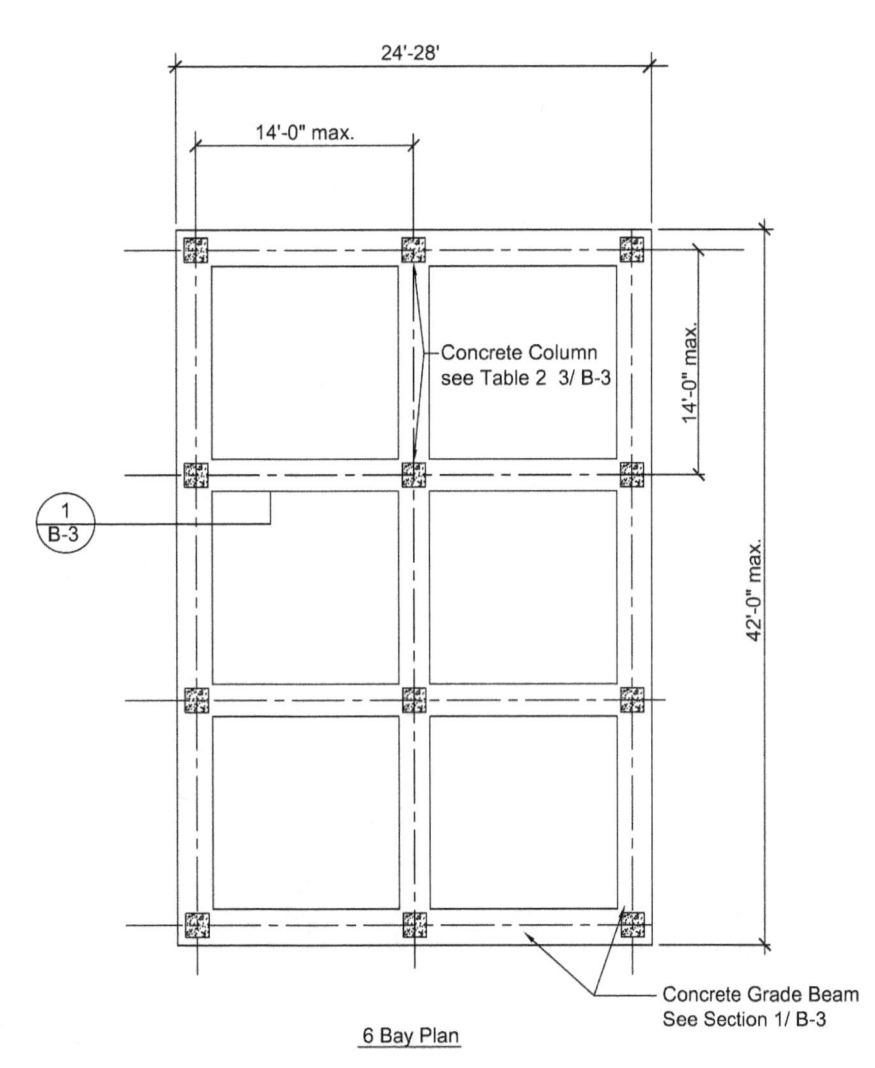

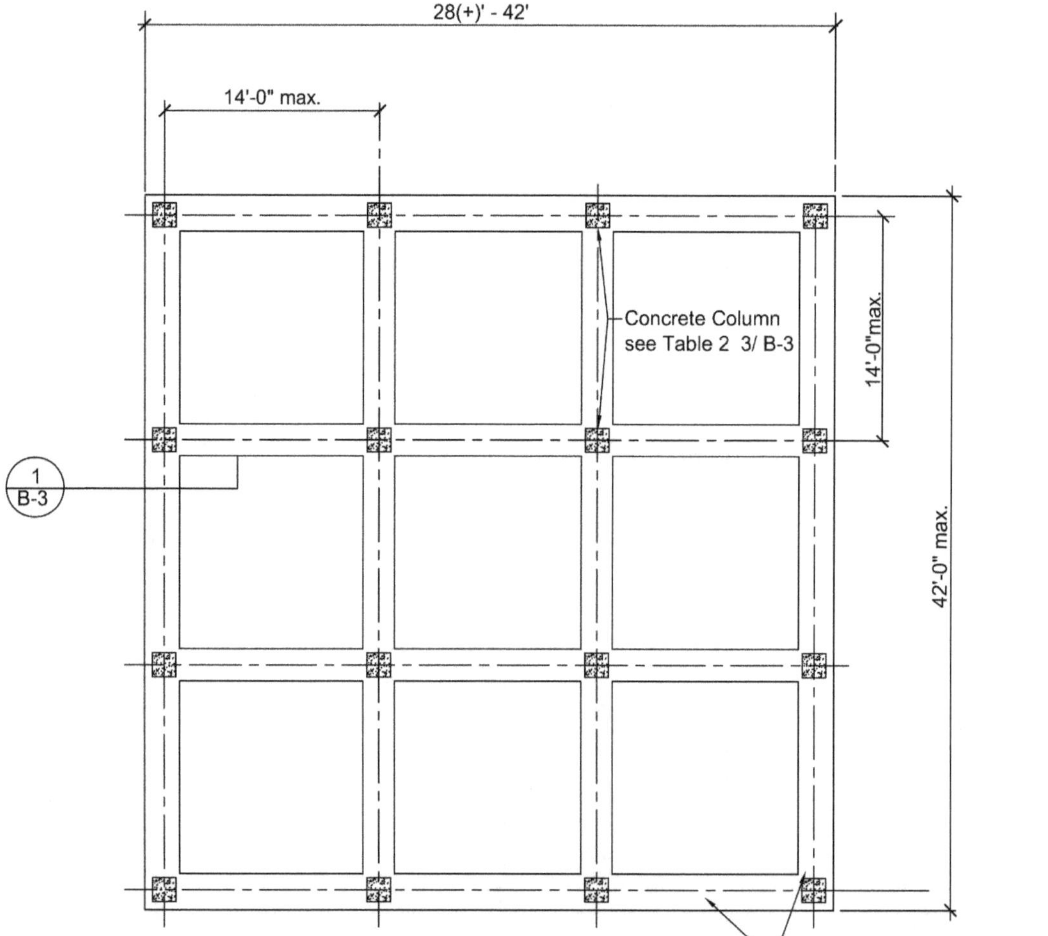

9 Bay Plan

Dimensions: 28(+)' – 42' overall; 14'-0" max. bay spacing; 42'-0" max. overall; 14'-0" max. bay spacing.

Labels: Concrete Column, see Table 2 3/B-3; Concrete Grade Beam, See Section 1/B-3; detail reference 1/B-3.

Case-B Open Foundation Steel Pipe Pile with Concrete Column and Grade Beam	
DRAWING NO.: B-2	SHEET 12 OF 31
DATE: August 8, 2006	
REVISED:	REV. 9

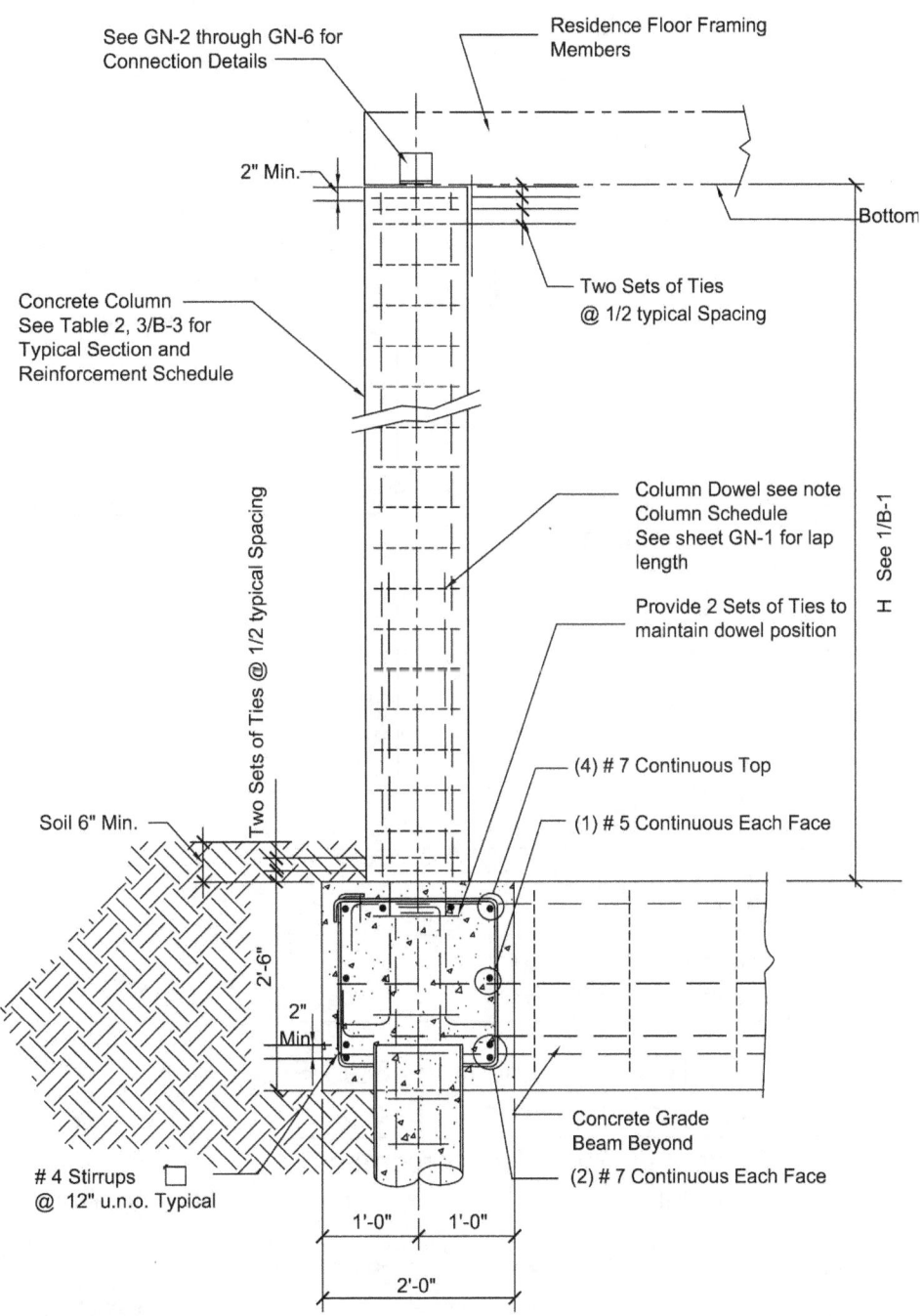

① Case B Open Foundation - Steel Pile/Concrete Column/Grade Beam
B-3 Column & Grade Beam Section
SCALE: NTS

**(2/B-3) Case B Open Foundation - Steel Pile/Concrete Column/Grade Beam
Typical Column Section**
SCALE: NTS

Size	16" x 16"	18" x 18"	20" x 20"
A	Verts (4) # 9 #4 Ties @ 16"		Verts (8) # 10 #4 Ties @ 18"
B	Verts (8) # 7 #4 Ties @ 14"	Verts (8) # 8 #4 Ties @ 16"	
C	Verts (8) # 8 #4 Ties @ 16"	Verts (8) # 9 #4 Ties @ 18"	
D	Verts (8) # 9 #4 Ties @ 16"	Verts (8) # 10 #4 Ties @ 18"	

Notes:
1) Provide same number and size of dowels as main reinforcing. See splice table on Sheet GN-1.

**(3/B-3) Case B Open Foundation - Steel Pile/Concrete Column/Grade Beam
Table 2 - Column Reinforcement Schedule**

Continuous reinforcing top and bottom. Use longest bar lengths practical. Place top bar splices at quarter point of span and bottom bar splices at columns. See splice table on Sheet GN-1.

Case-B Open Foundation
Steel Pipe Pile with Concrete Column and Grade Beam

DRAWING NO.: B-3	SHEET 13 OF 31
DATE: August 8, 2006	
REVISED:	REV. 9

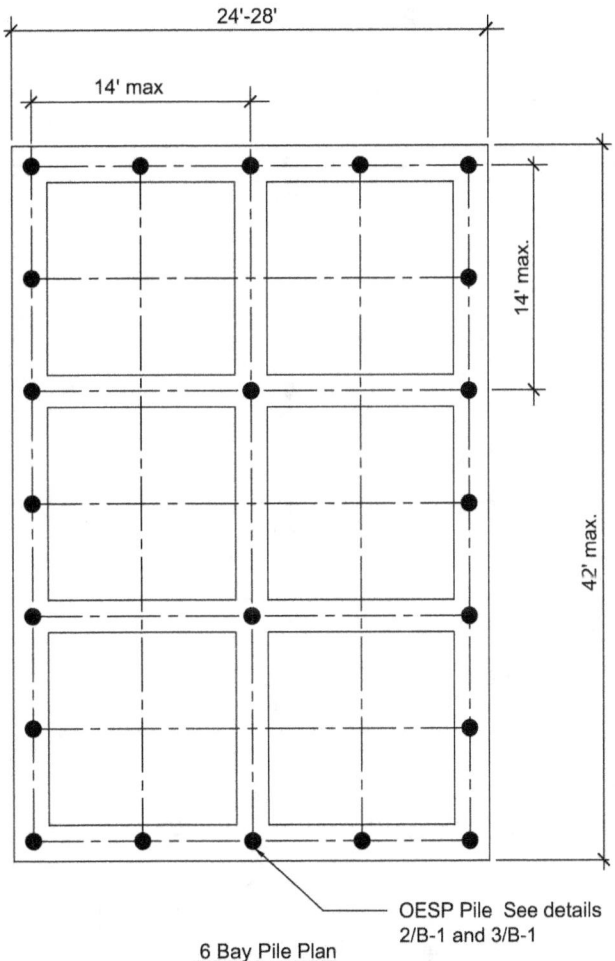

6 Bay Pile Plan

One Story and Two Story Residences Height H up to 12' - for winds up to 150 mph

Case B Open Foundation - Steel Pile/ Concrete Column/Grade Beam

1/B-4   22 Piles
SCALE: NTS

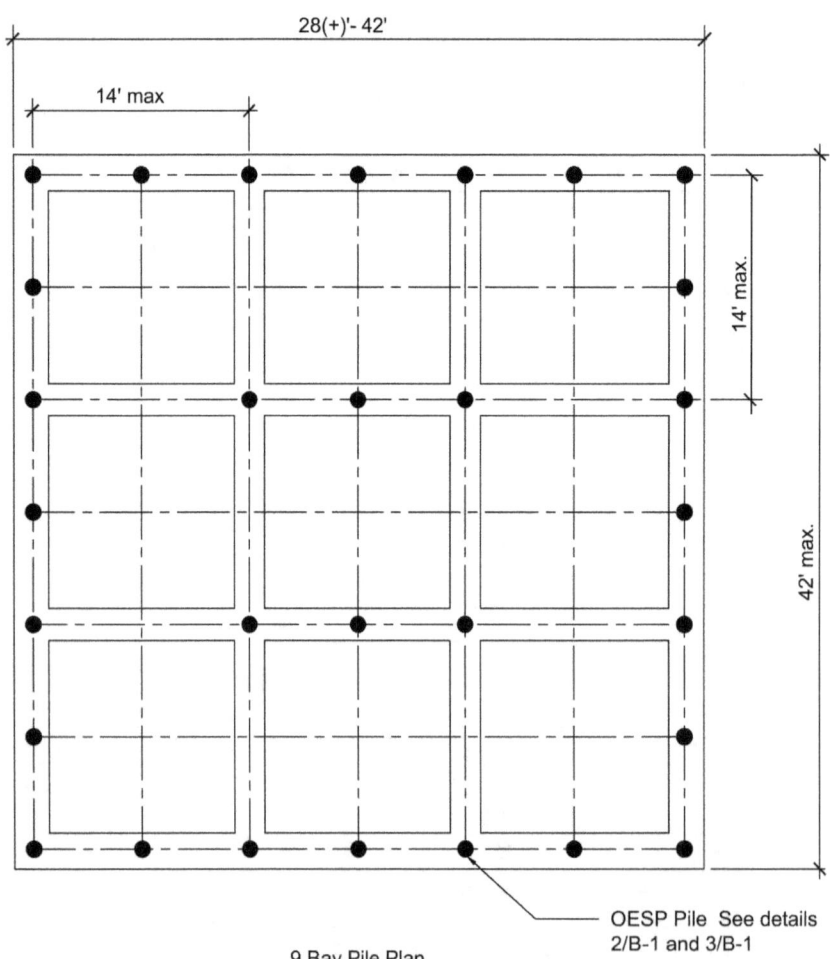

9 Bay Pile Plan

One Story and Two Story Residences Height H up to 15' - for winds up to 150 mph

**Case-B Open Foundation
Steel Pipe Pile with Concrete Column and Grade Beam**

DRAWING NO.: B-4	SHEET 14 OF 31
DATE: August 8, 2006	
REVISED:	REV. 9

Case B Open Foundation - Steel Pile/ Concrete Column/Grade Beam

2/B-4 30 Piles
SCALE: NTS

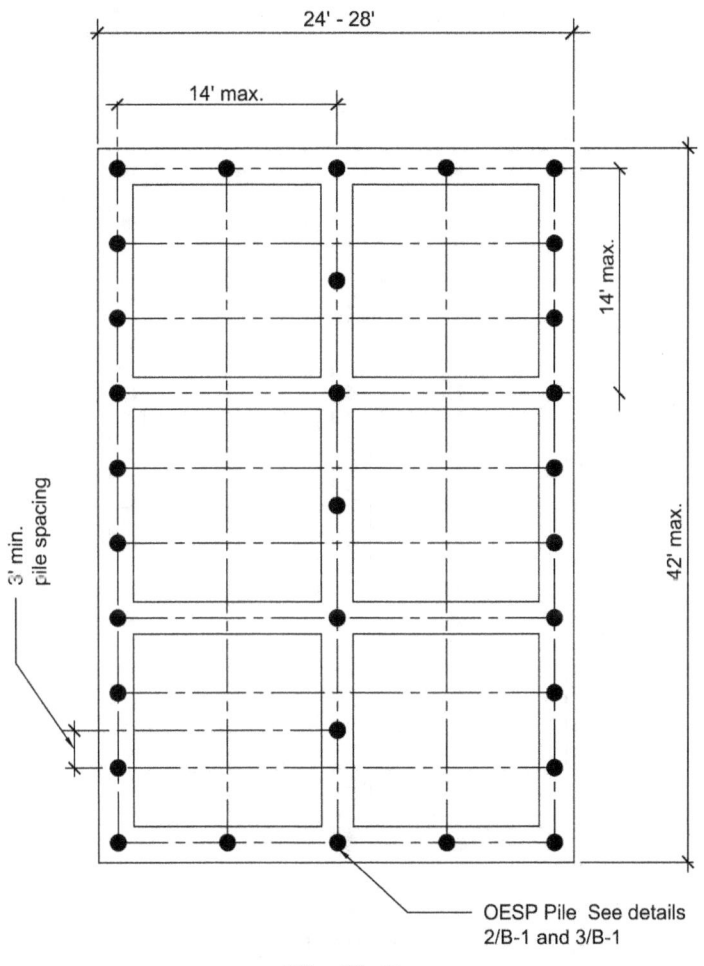

6 Bay Pile Plan
One Story and Two Story Residences Height H up to 12' - for winds up to 150 mph

Case B Open Foundation - Steel Pile/ Concrete Column/Grade Beam

1 / B-5
31 PILES
SCALE: NTS

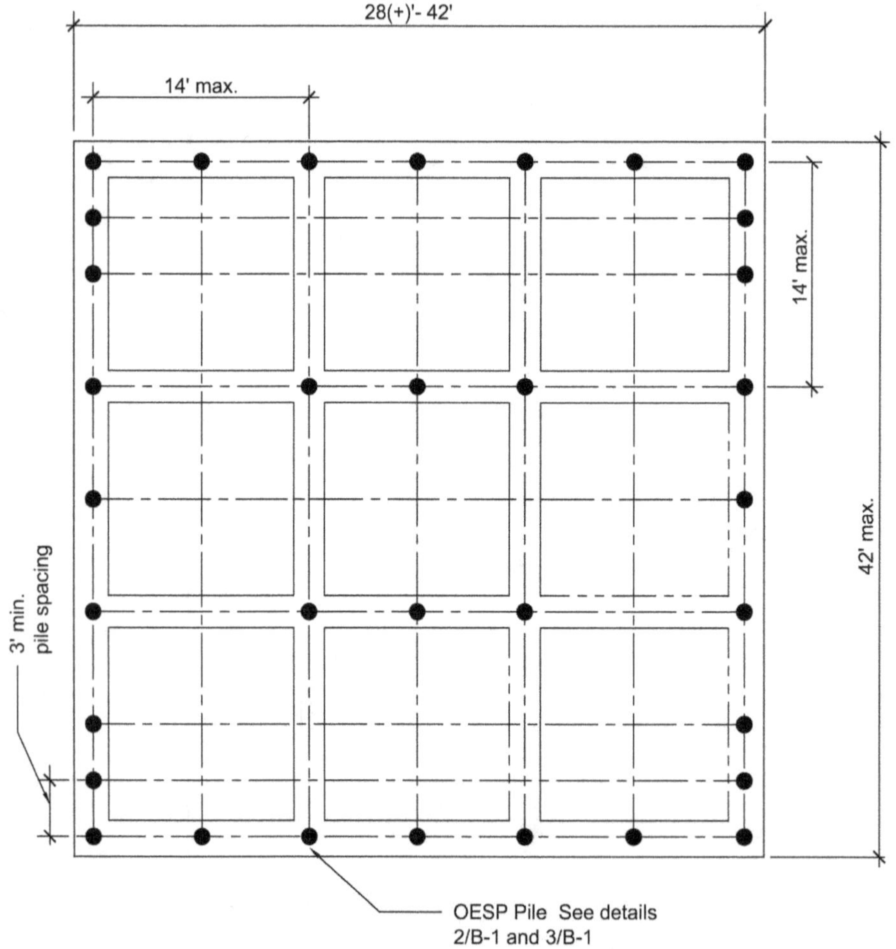

9 Bay Pile Plan

One Story and Two Story Residences Height H up to 15' - for winds up to 150 mph

Case B Open Foundation - Steel Pile/Concrete Column/Grade Beam
2/B-5
36 PILES
SCALE: NTS

Case-B Open Foundation
Steel Pipe Pile with Concrete Column and Grade Beam

DRAWING NO.: B-5	SHEET 15 OF 31
DATE: August 8, 2006	
REVISED:	REV. 9

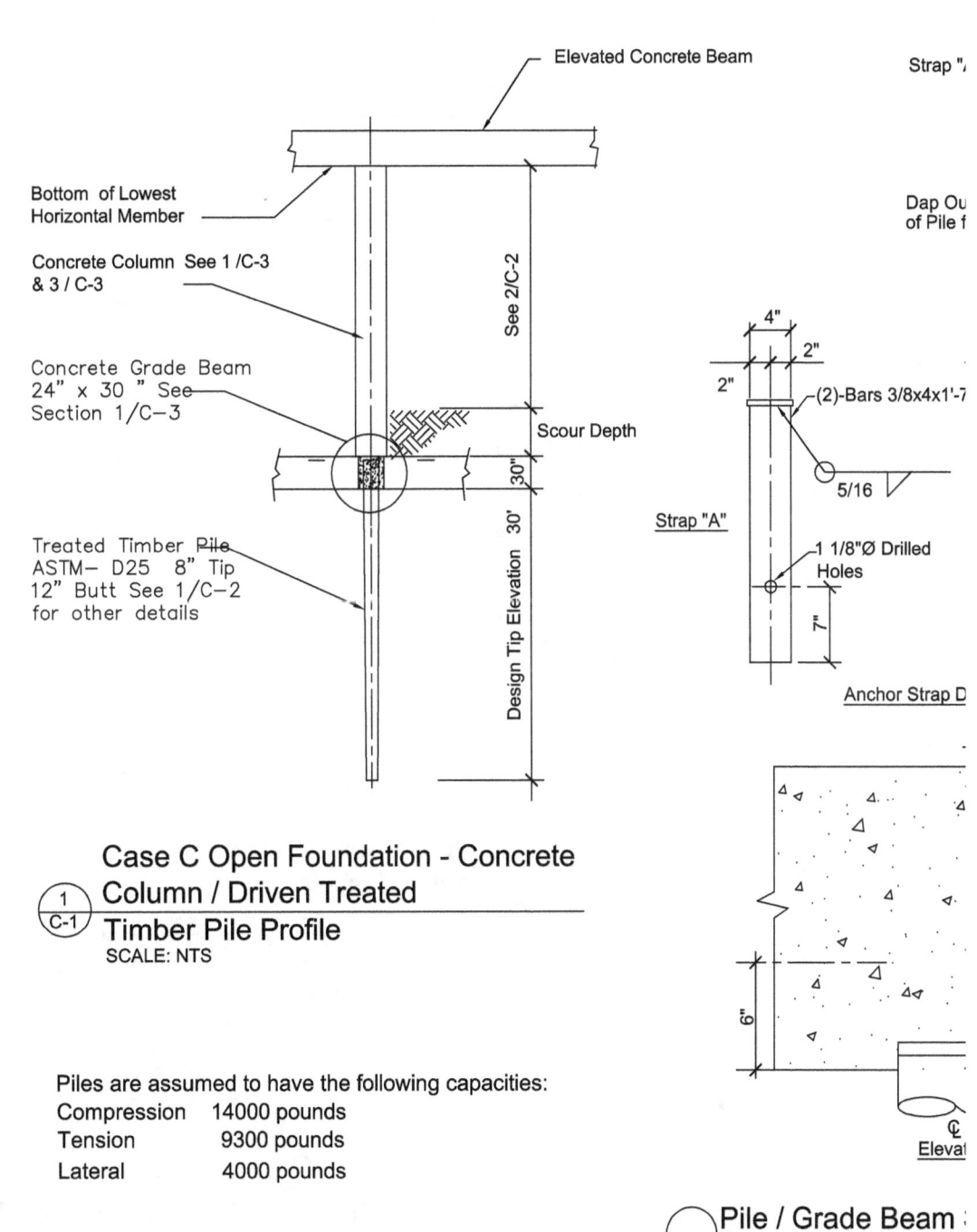

Case C Open Foundation - Concrete Column / Driven Treated Timber Pile Profile
1/C-1 SCALE: NTS

Piles are assumed to have the following capacities:
Compression   14000 pounds
Tension   9300 pounds
Lateral   4000 pounds

Pile / Grade Beam
Pile Detail
SCALE: NTS

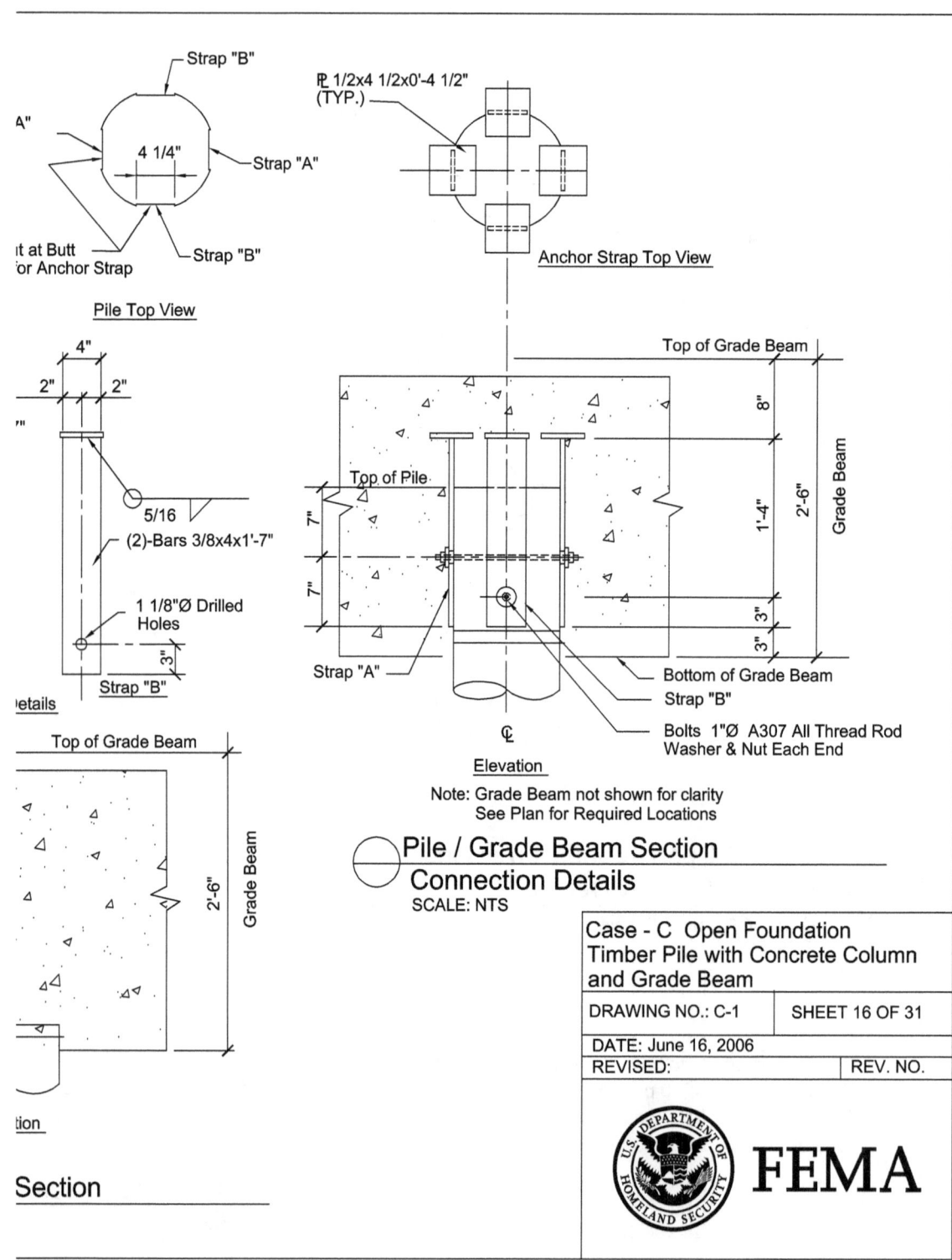

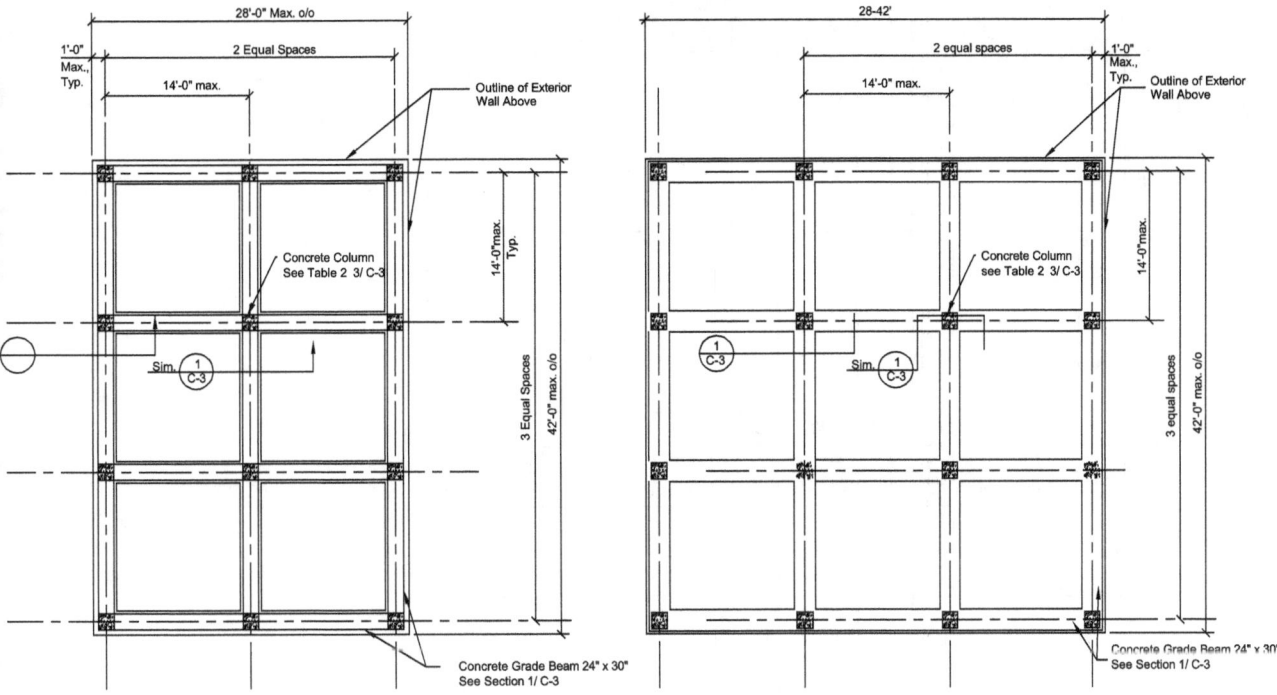

6 - Bay Foundation Column and Beam Plan

9 - Bay Foundation Column and Beam Plan

Case C - Open Foundation - Column & Grade Beam Plan - 1
SCALE: NTS

Case C - Open Foundation - Column & Grade Beam Plan - 2
SCALE: NTS

| Column Size & Reinforcement Schedule - One Story ||||||||
| "H" | 150 mph || 140 mph || 130 mph || 120 mph |
	Size (sq)	Reinforcing	Size (sq)	Reinforcing	Size (sq)	Reinforcing	Size (sq)	Reinforcing
8 ft	16"	A	16"	A	16"	A	16"	A
10 ft	16"	C	16"	C	16"	B	16"	B
12 ft	18"	D	18"	D	16"	D	16"	D
15 ft	20"	A	20"	A	18"	D	18"	C

| Column Size & Reinforcement Schedule - Two Story ||||||||
| "H" | 150 mph || 140 mph || 130 mph || 120 mph |
	Size (sq)	Reinforcing	Size (sq)	Reinforcing	Size (sq)	Reinforcing	Size (sq)	Reinforcing
8 ft	16"	B	16"	B	16"	B	16"	A
10 ft	16"	D	16"	C	16"	C	16"	B
12 ft	18"	D	18"	C	18"	C	18"	B
15 ft	20"	A	20"	A	18"	D	18"	D

(3/C-2) **Case B - Table 1 Column Size & Reinforcement Schedule**

Notes:
1) See Table 2, 3/B3 sheet for Column size and reinforcement details

Case- C Open Foundation Concrete Timber Pile with Concrete Column and Grade Beam

DRAWING NO.: C-2	SHEET 17 OF 31
DATE: June 16, 2006	
REVISED:	REV. NO.

FEMA

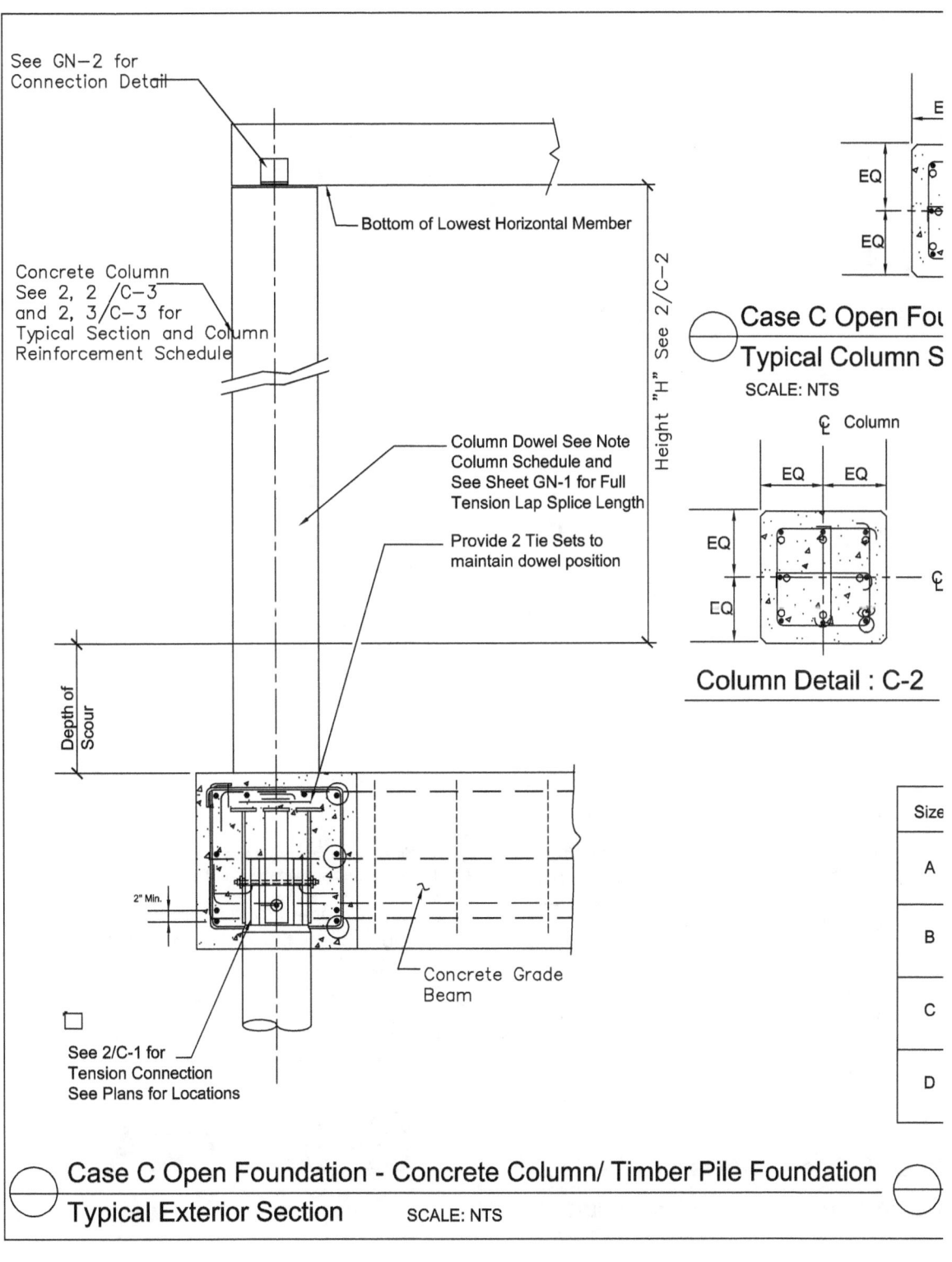

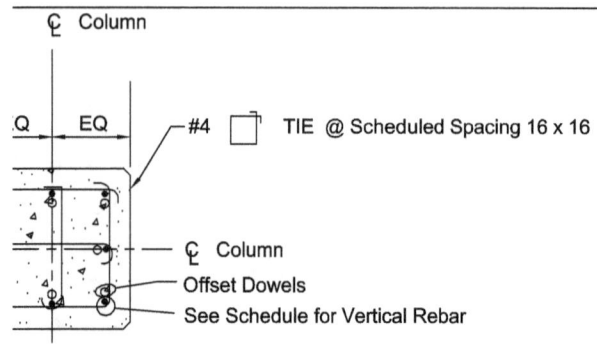

Foundation - Concrete Column Sections
Section

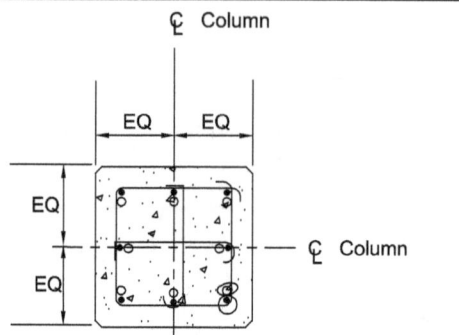

Column Detail : C-1

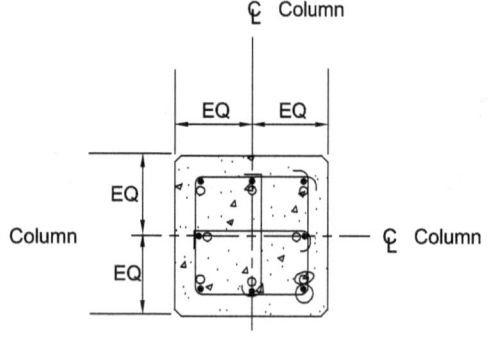

Column Detail : C-3

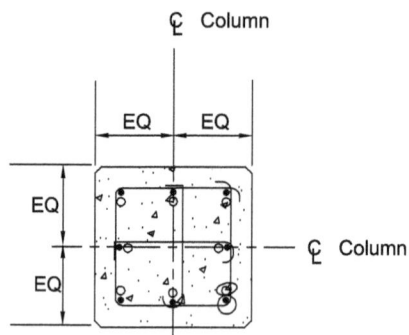

Column Detail : C-4

	16" x 16"	18" x 18"	20" x 20"
	Verts (4) # 9 #4 Ties @ 16"	Verts (8) # 7 #4 Ties @ 14"	Verts (8) # 10 #4 Ties @ 18"
	Verts (8) # 7 #4 Ties @ 14"	Verts (8) # 8 #4 Ties @ 16"	
	Verts (8) # 8 #4 Ties @ 16"	Verts (8) # 9 #4 Ties @ 18"	
	Verts (8) # 9 #4 Ties @ 16"	Verts (8) # 10 #4 Ties @ 18"	

Case C Open Foundation - Table 2
Column Reinforcement Schedule

Notes:
1) Provide same number and size of dowels as main reinforcing. See splice table on Sheet GN-1.

Case- C Open Foundation Concrete Timber Pile with Concrete Column and Grade Beam

DRAWING NO.: C-3   SHEET 18 OF 31
DATE: June 16, 2006
REVISED: REV. 1

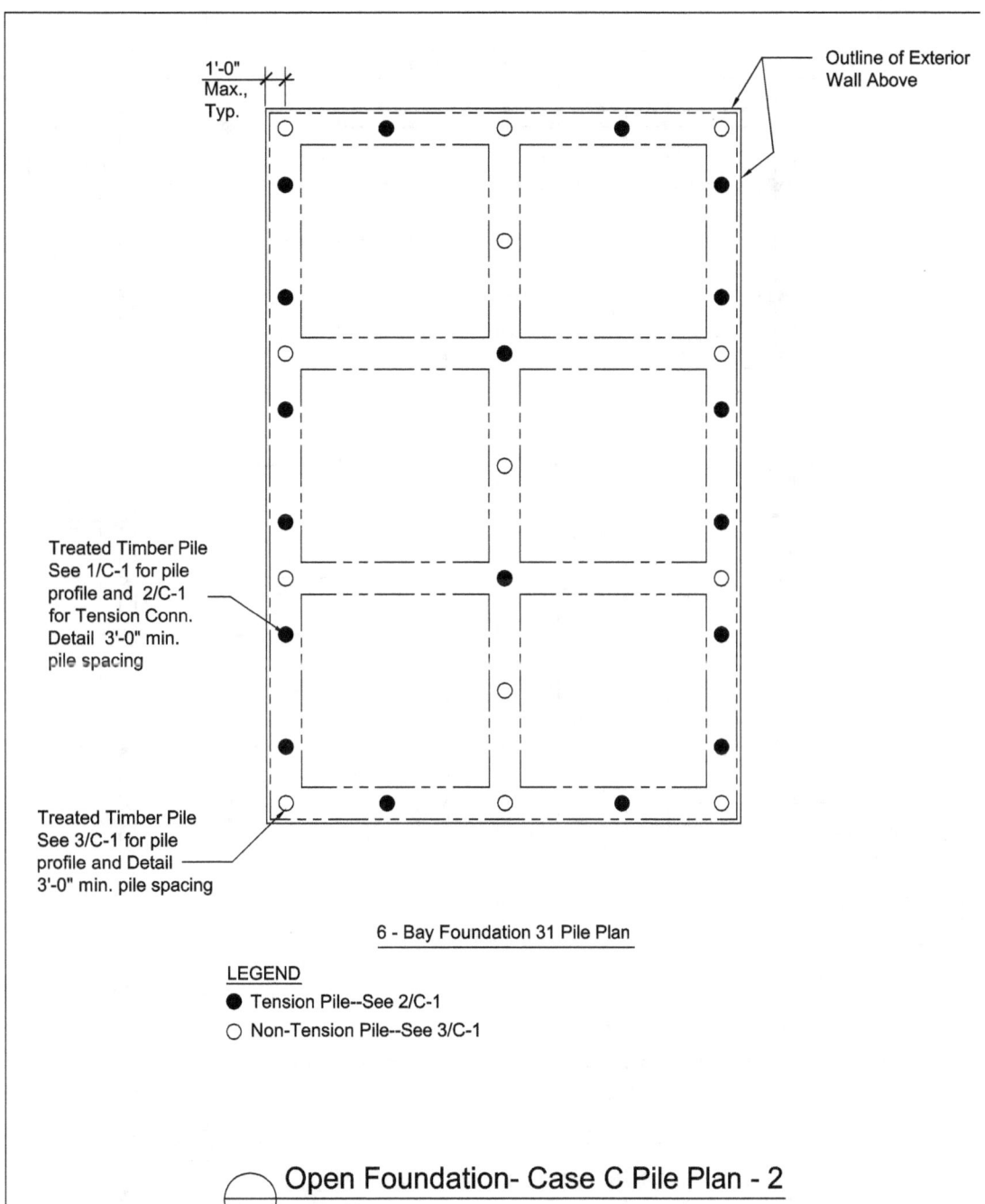

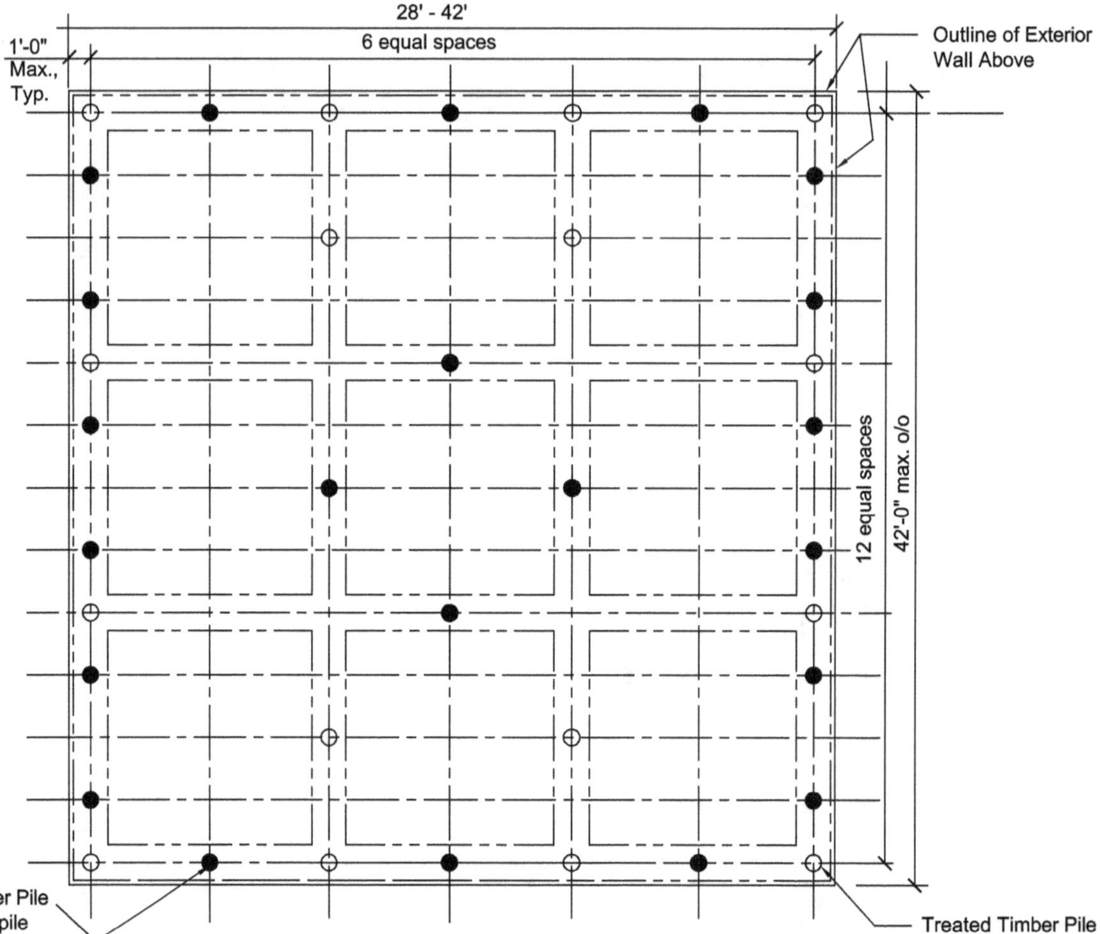

9 Bay Pile Plan

Dimensions: 28' - 42', 6 equal spaces; 42'-0" max. o/o, 12 equal spaces; 1'-0" Max., Typ.

Outline of Exterior Wall Above

Treated Timber Pile See 1/C-1 for pile profile and 2/C-1 for Tension Conn. Detail 3'-0" min. pile spacing

Treated Timber Pile See 1/C-1 for pile profile and Detail 3'-0" min. pile spacing

LEGEND
- ● Tension Pile--See 2/C-1
- ○ Non-Tension Pile--See 3/C-1

One and Two Residences Height H up to 10' - for winds up to 140 mph

2/C-4 Open Foundation- Case C Pile Plan - 2
38 PILES
SCALE: NTS

Case- C Open Foundation Concrete Timber Pile with Concrete Column and Grade Beam

DRAWING NO.: C-4	SHEET 19 OF 31
DATE: June 16, 2006	
REVISED:	REV. NO.

FEMA

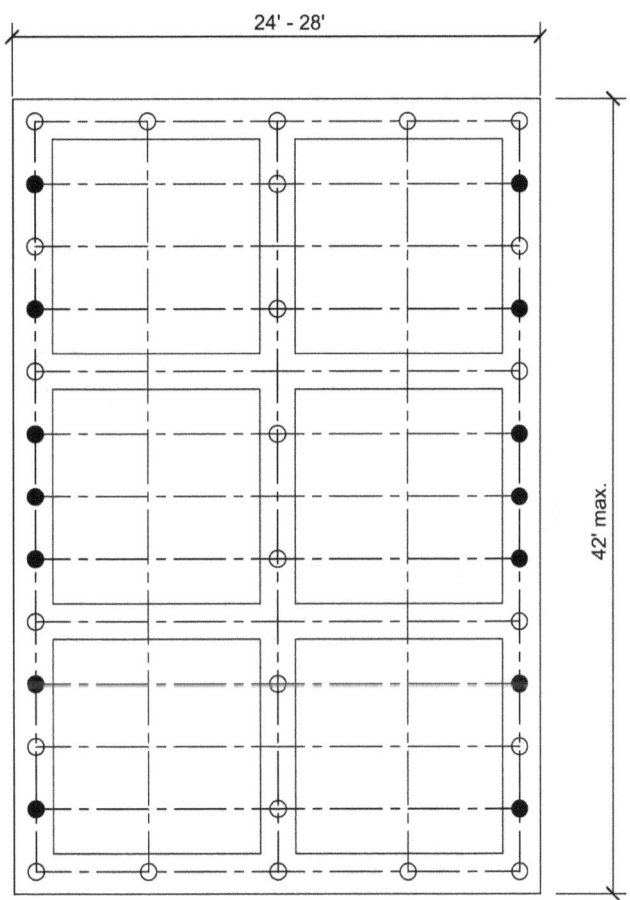

6 Bay Pile Plan

LEGEND
- ● Tension Pile--See 2/C-1
- ○ Non-Tension Pile--See 3/C-1

One and Two Story Residences Height H up to 10' - for winds up to 150 mph

Case C Open Foundation - Timber Pile with Concrete Column & Grade Beam

1/C-5 Pile Plan
38 PILES
SCALE: NTS

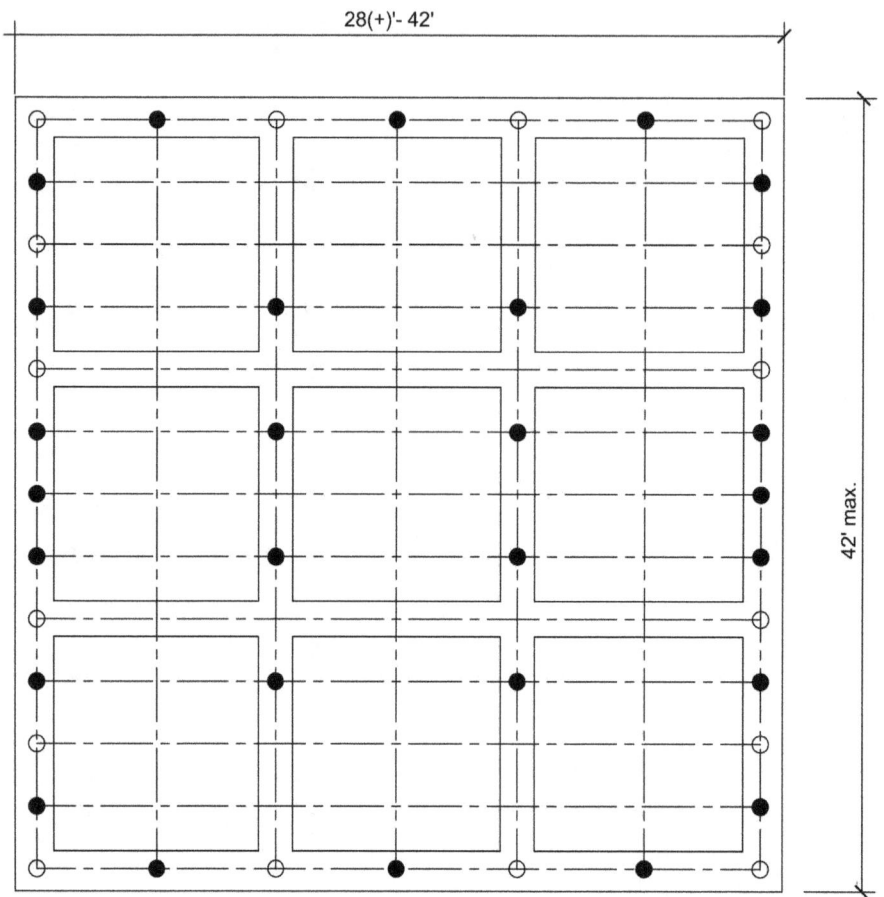

9 Bay Pile Plan

LEGEND
● Tension Pile--See 2/C-1
○ Non-Tension Pile--See 3/C-1

One and Two Story Residences Height H up to 10' - for winds up to 150 mph

Case C Open Foundation - Timber Pile with Concrete Column & Grade Beam

(2/C-5) **Pile Plan**
44 PILES
SCALE: NTS

Case- C Open Foundation Timber Pile with Concrete Column and Grade Beam	
DRAWING NO.: C-5	SHEET 20 OF 31
DATE: August 8, 2006	
REVISED:	REV. 9

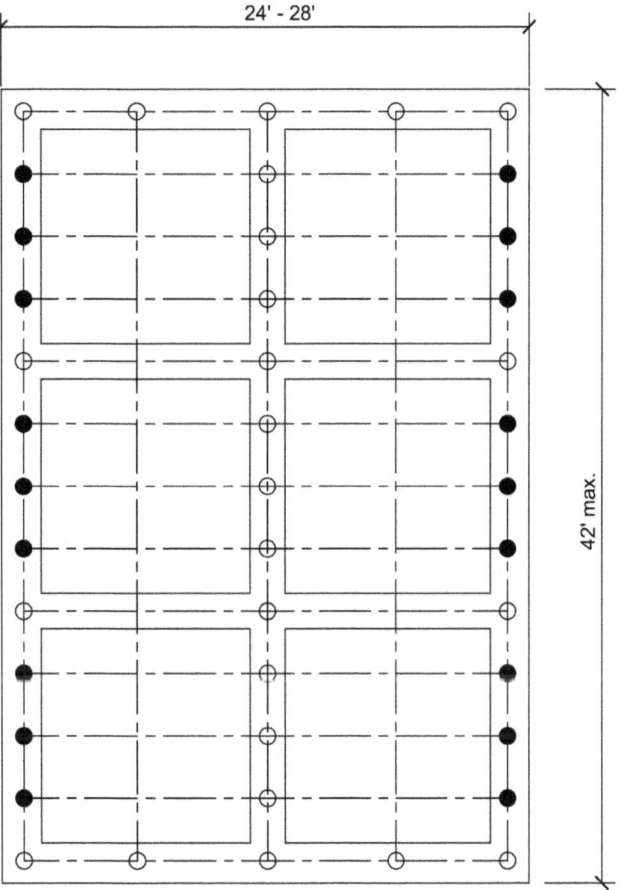

6 Bay Pile Plan

LEGEND
● Tension Pile--See 2/C-1
○ Non-Tension Pile--See 3/C-1

One Story & Two Story Residences Height H
up to 15' - for winds up to 150 mph

Case C Open Foundation - Timber Pile with Concrete Column & Grade Beam
1/C-6 Pile Plan
43 PILES
SCALE: NTS

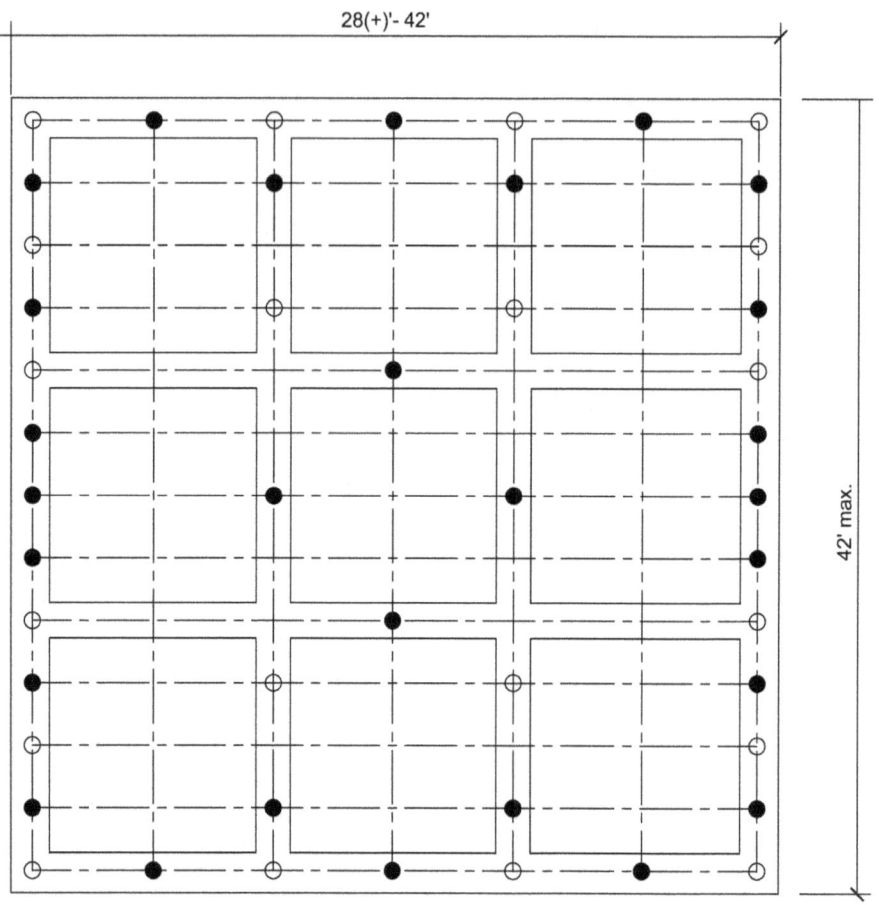

9 Bay Pile Plan

LEGEND
- ● Tension Pile--See 2/C-1
- ○ Non-Tension Pile--See 3/C-1

One Story & Two Story Residences Height H up to 15' - for winds up to 150 mph

Case C Open Foundation - Timber Pile with Concrete Column & Grade Beam

(2/C-6) Pile Plan
48 PILES
SCALE: NTS

Case- C Open Foundation Timber Pile with Concrete Column and Grade Beam	
DRAWING NO.: C-6	SHEET 21 OF 31
DATE: August 8, 2006	
REVISED:	REV. 9

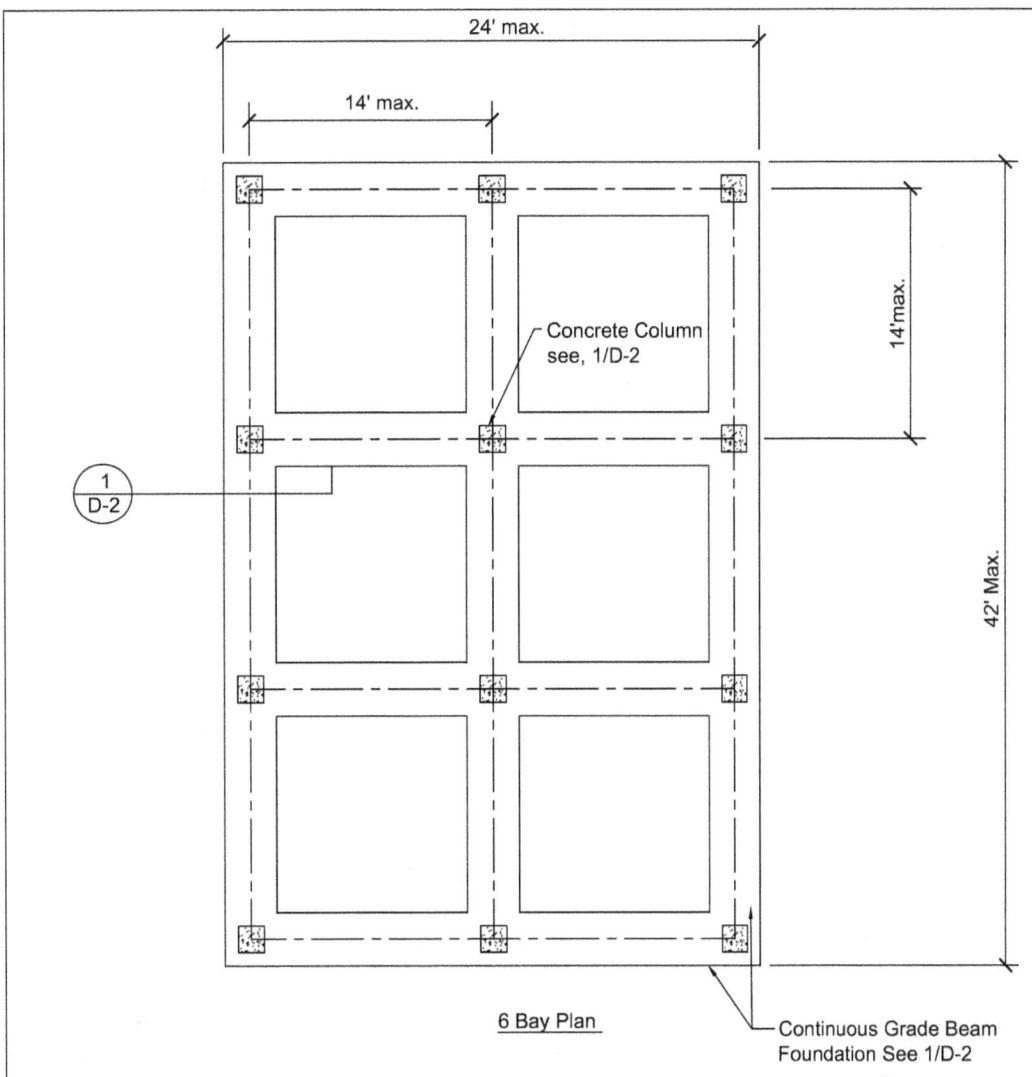

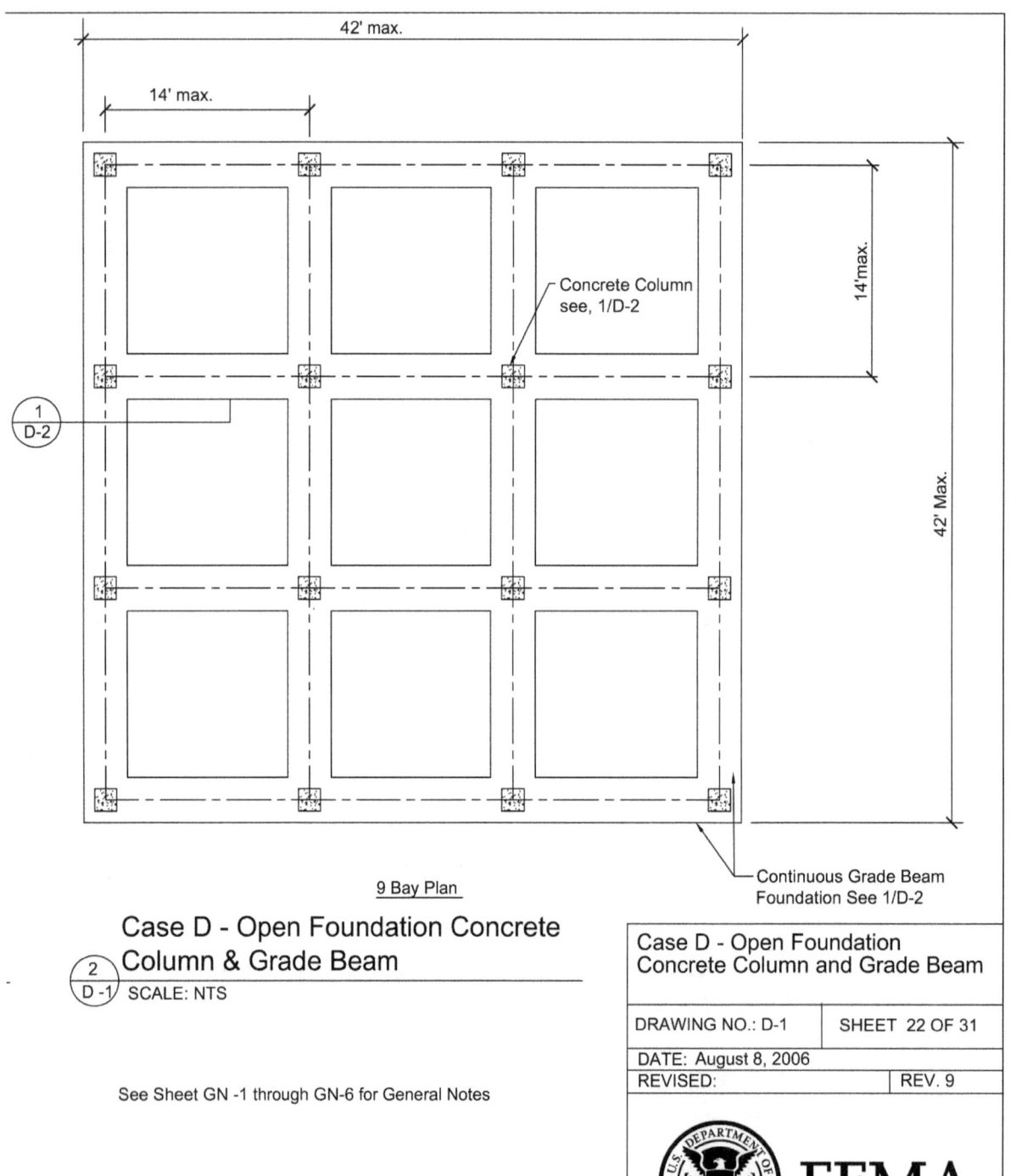

9 Bay Plan

Case D - Open Foundation Concrete Column & Grade Beam
2/D-1 SCALE: NTS

See Sheet GN-1 through GN-6 for General Notes

Case D - Open Foundation Concrete Column and Grade Beam	
DRAWING NO.: D-1	SHEET 22 OF 31
DATE: August 8, 2006	
REVISED:	REV. 9

FEMA

Wind Speed (mph)				
One Story				
Height H	150	140	130	120
5'	2'-6" x 2'-0" (4) - # 6 Column 16 x 16 A	2'-3" x 2'-0" (4) - # 6 Column 16 x 16 A	2'-0" x 2'-0" (4) - # 6 Column 16 x 16 A	2'-0" x 2'-0" (4) - # 6 Column 16 x 16 A
6'	2'-9" x 2'-0" (5) - # 6 Column 16 x 16 A	2'-6" x 2'-0" (4) - # 6 Column 16 x 16 A	2'-3" x 2'-0" (4) - # 6 Column 16 x 16 A	2'-0" x 2'-0" (4) - # 6 Column 16 x 16 A
8'	3'-3" x 2'-0" (5) - # 6 Column 16 x 16 B	2'-6" x 2'-0" (4) - # 6 Column 16 x 16 B	2'-3" x 2'-0" (4) - # 6 Column 16 x 16 A	2'-0" x 2'-0" (4) - # 6 Column 16 x 16 A
Two Story				
5'	3'-6" x 2'-3" (6) - # 6 Column 16 x 16 A	2'-9" x 2'-3" (5) - # 6 Column 16 x 16 A	2'-6" x 2'-0" (5) - # 6 Column 16 x 16 A	2'-6" x 2'-0" (4) - # 6 Column 16 x 16 A
6'	4'-0" x 2'-3" (5) - # 7 Column 16 x 16 A	3'-0" x 2'-3" (6) - # 6 Column 16 x 16 A	3'-0" x 2'-0" (6) - # 6 Column 16 x 16 A	2'-6" x 2'-0" (4) - # 6 Column 16 x 16 A
8'	4'-0" x 2'-3" (5) - # 7 Column 16 x 16 B	3'-3" x 2'-3" (5) - # 6 Column 16 x 16 B	3'-0" x 2'-0" (5) - # 6 Column 16 x 16 A	2'-6" x 2'-0" (5) - # 6 Column 16 x 16 A
Height H	150	140	130	120
Wind Speed (mph)				

1/D-2 Case D - Open Foundation Concrete Column & Grade Beam
Table 1 Continuous Grade Beam Size and Concrete Column Schedule

Notes:
1) Continuous reinforcing top and bottom. Use longest bar lengths that are practical. Place top bar splices at quarter point of span and bottom bar splices at columns.
2) See Table 3/D3 for reinforcement schedule.
3) Legend: 2'-6"x2'-0"   - Footing Size A (width) x B (height)
                  (4) - #6   - Footing Rebar, top & bottom
                  Column 16 x 16 B   - Column Size; see 1/D-2

Case D - Open Foundation
Concrete Column and Grade Beam

DRAWING NO.: D-2	SHEET 23 OF 31
DATE: August 8, 2006	
REVISED:	REV. 9

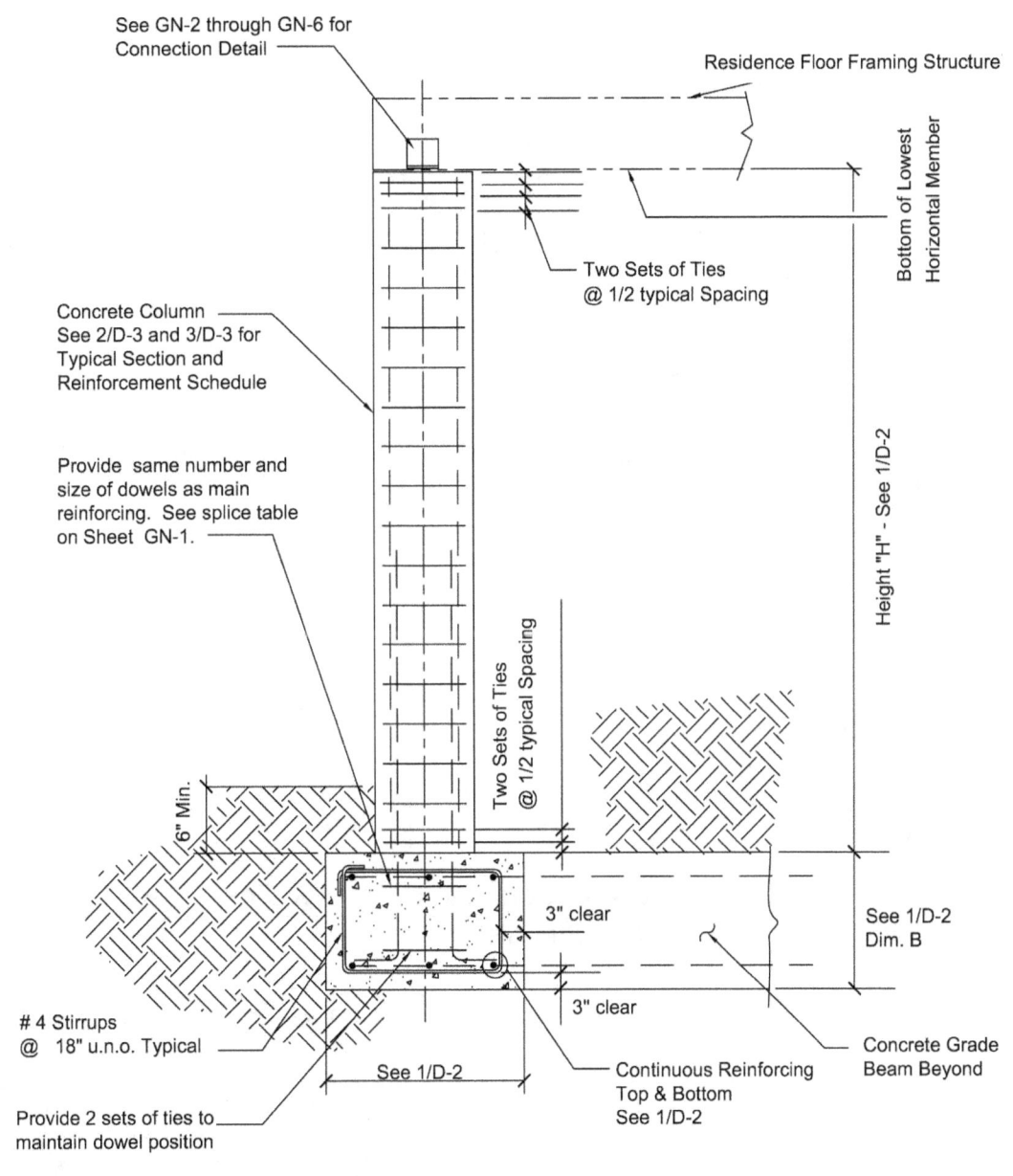

1/D-3 Case D Open Foundation - Concrete Column & Grade Beam
Typical Exterior Section
SCALE: NTS

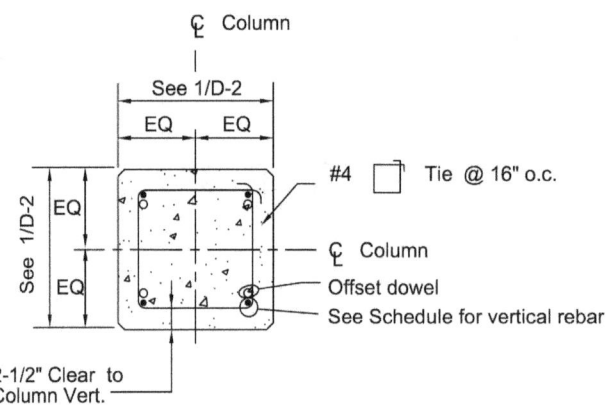

2/D-3 Case D Open Foundation - Concrete Column & Grade Beam
Typical Column Section
SCALE: NTS

Size	16" x 16"
A	Verts (4) # 8 #4 Ties @ 16"
B	Verts (4) # 9 #4 Ties @ 16"

3/D-3 Case D Open Foundation - Concrete Column & Grade Beam
Column Reinforcement Schedule

Case D - Open Foundation
Concrete Column and Grade Beam

DRAWING NO.: D-3	SHEET 24 OF 31
DATE: August 8, 2006	
REVISED:	REV. 9

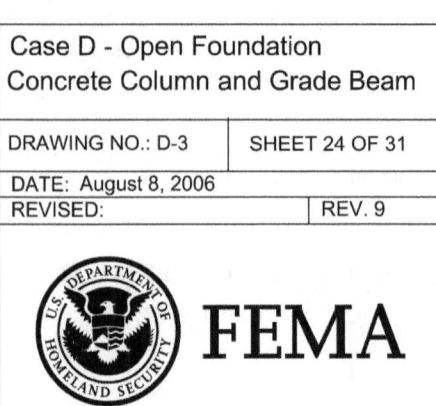

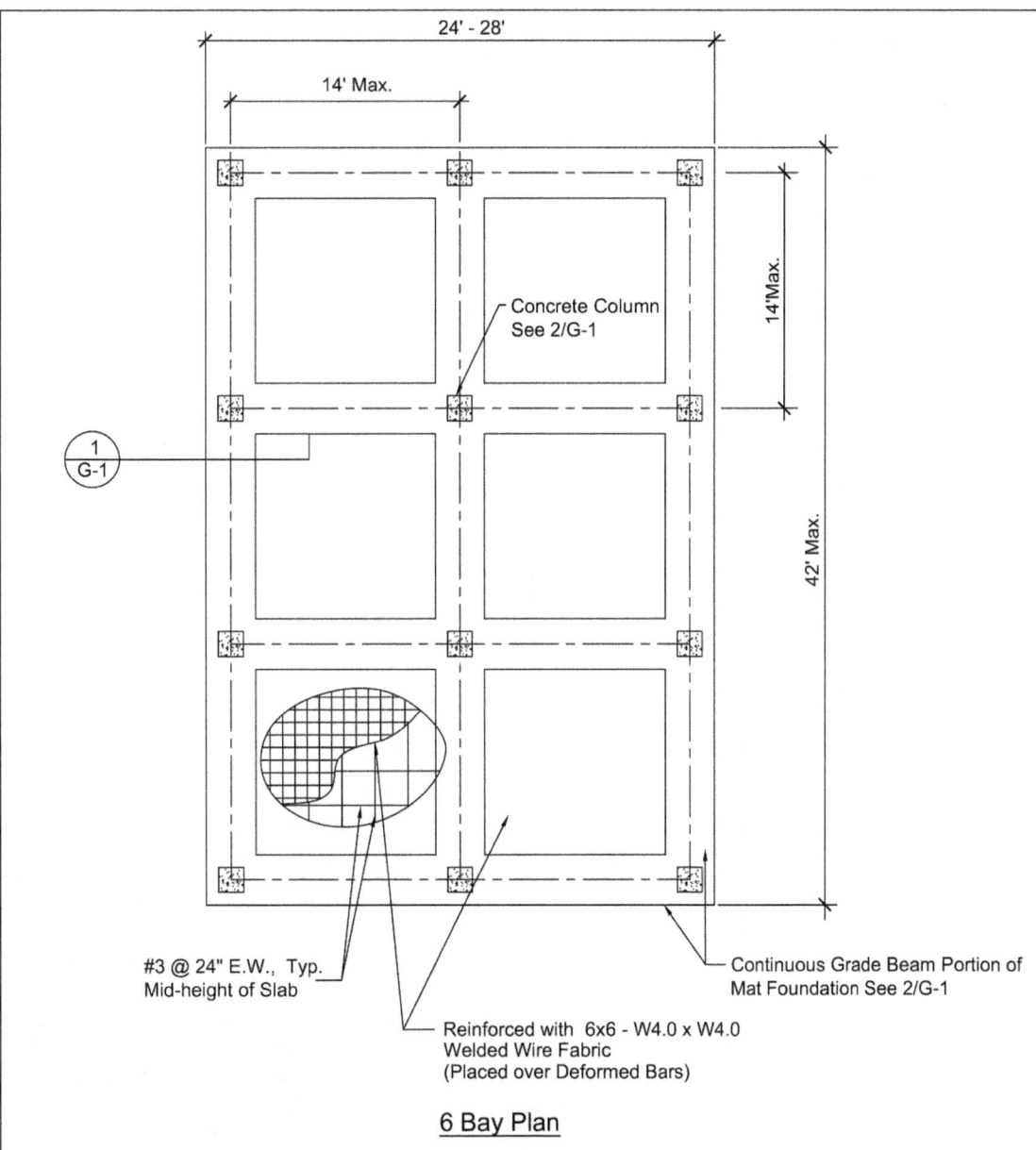

6 Bay Plan

Case G - Open Foundation Concrete Column & Grade Beam with Slab
1/G-1 Plan
SCALE: NTS

Notes:
1) Continuous reinforcing top and bottom. Use longest bar lengths that are practical. Place top bar splices at quarter point of span and bottom bar splices at columns.
2) See 3/G-2 for reinforcement details.

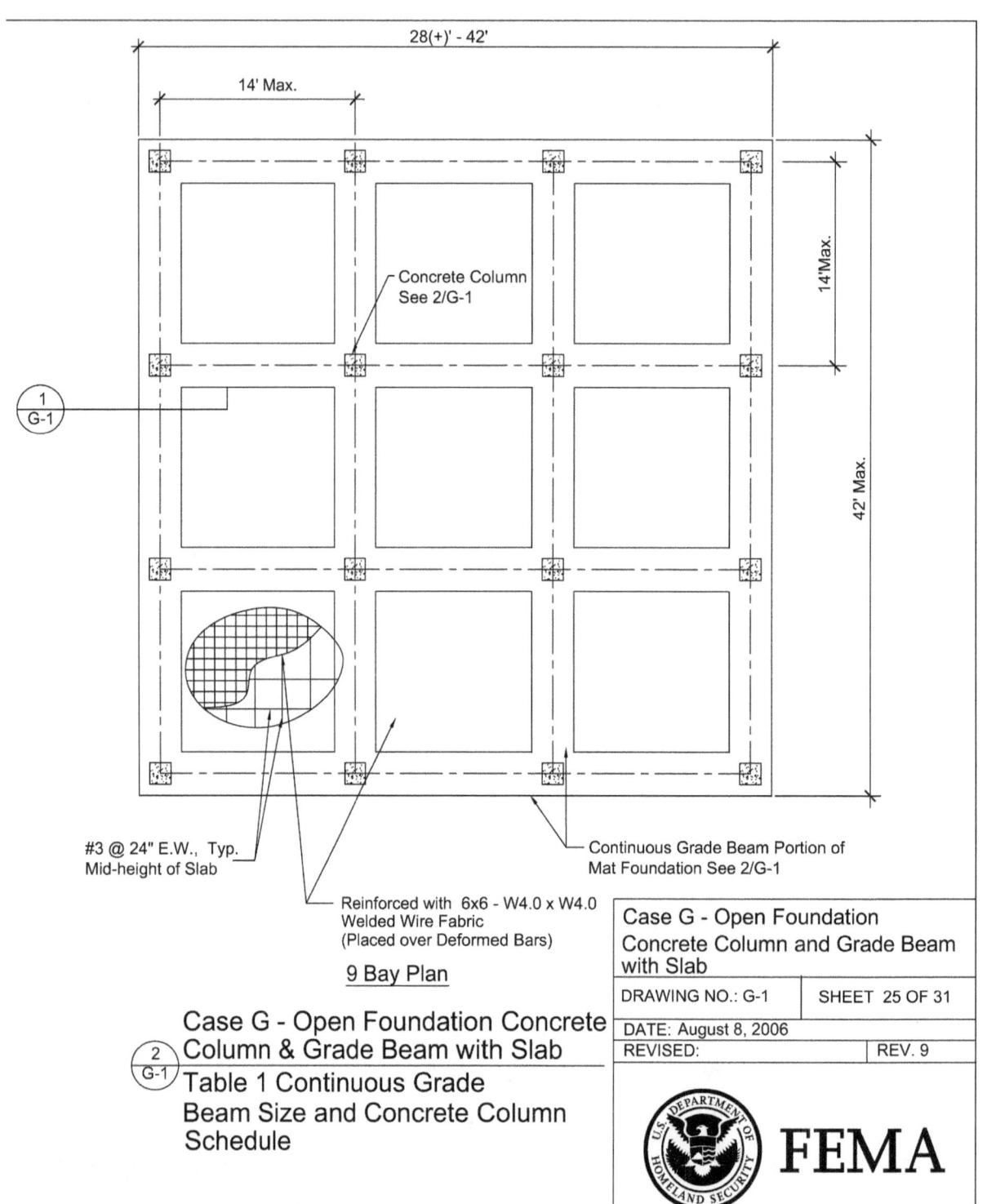

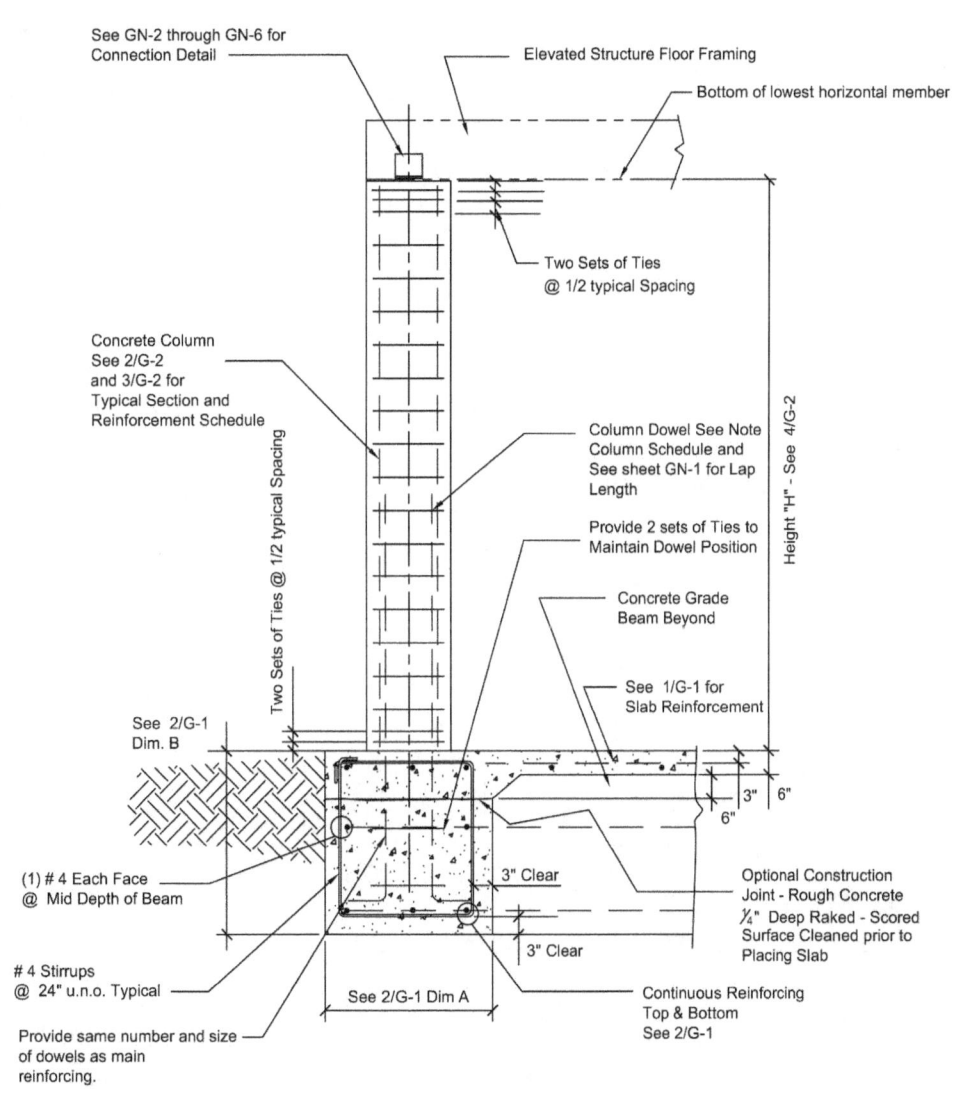

Case G - Open Foundation Concrete Column & Grade Beam with Slab
Typical Exterior Section
SCALE: NTS
(1/G-2)

Case G - Open Foundation Concrete Column & Grade Beam with Slab
Table 1 Continuous Grade Beam Size and Concrete Column Schedule
(4/G-2)

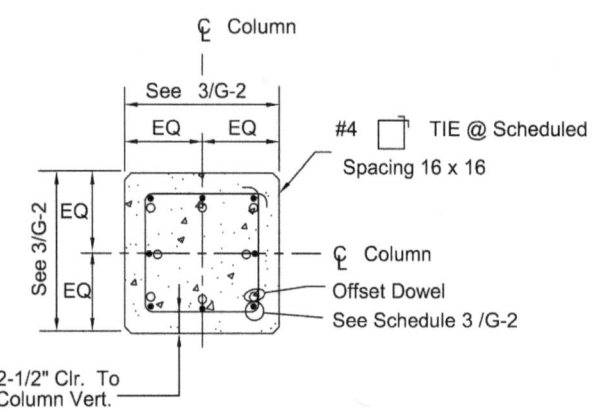

Case G Open Foundation Concrete Column & Grade Beam
2/G-2 Typical Column Section
SCALE: NTS

Case G - Open Foundation Column and Grade Beam
3/G-2 Column Reinforcement Schedule

Size	16 16" x 16"
A	Verts (4) # 8 #4 Ties @ 16"
B	Verts (8) # 7 #4 Ties @ 14"
C	Verts (8) # 8 #4 Ties @ 14"

Legend:
2'-6"x4'   - Footing Size A (width) x B (height)
(3) - #6   - Footing Rebar, Top & Bottom
Column 16 x 16 A   - Column Size; see 3/G-2

	Wind Speed mph			
	One Story			
Height H	150	140	130	120
8'	2'-6" x 4' (4) - # 6 Column 16 x 16 A	2' x 4' (3) - # 6 Column 16 x 16 A	2' x 4' (3) - # 6 Column 16 x 16 A	2' x 4' (3) - # 6 Column 16 x 16 A
10'	2'-9" x 4' (4) - # 6 Column 16 x 16 A	2' x 4' (3) - # 6 Column 16 x 16 A	2' x 4' (3) - # 6 Column 16 x 16 A	2' x 4' (3) - # 6 Column 16 x 16 A
12'	3' x 4' (4) - # 6 Column 16 x 16 B	2'-3" x 4' (3) - # 6 Column 16 x 16 B	2' x 4' (3) - # 6 Column 16 x 16 A	2' x 4' (3) - # 6 Column 16 x 16 A
15'	3'-6" x 4' (5) - # 6 Column 16 x 16 C	2'-9" x 4' (4) - # 6 Column 16 x 16 B	2'-3" x 4' (3) - # 6 Column 16 x 16 B	2' x 4' (3) - # 6 Column 16 x 16 B
	Two Story			
8'	3' x 4' (4) - # 6 Column 16 x 16 B	3' x 4' (4) - # 6 Column 16 x 16 A	2'-3" x 4' (3) - # 6 Column 16 x 16 A	2' x 4' (3) - # 6 Column 16 x 16 A
10'	4'-3" x 4' (5) - # 6 Column 16 x 16 B	3'-6" x 4' (5) - # 6 Column 16 x 16 B	2'-6" x 4' (4) - # 6 Column 16 x 16 B	2' x 4' (3) - # 6 Column 16 x 16 B
12'	4'-6" x 4' (4) - # 7 Column 16 x 16 C	4' x 4' (5) - # 6 Column 16 x 16 C	2'-9" x 4' (4) - # 7 Column 16 x 16 B	2'-3" x 4' (3) - # 6 Column 16 x 16 B
Height H	150	140	130	120
	Wind Speed mph			

Case G - Open Foundation Concrete Column and Grade Beam with Slab

DRAWING NO.: G-2   SHEET 26 OF 31
DATE: August 8, 2006
REVISED: REV. 9

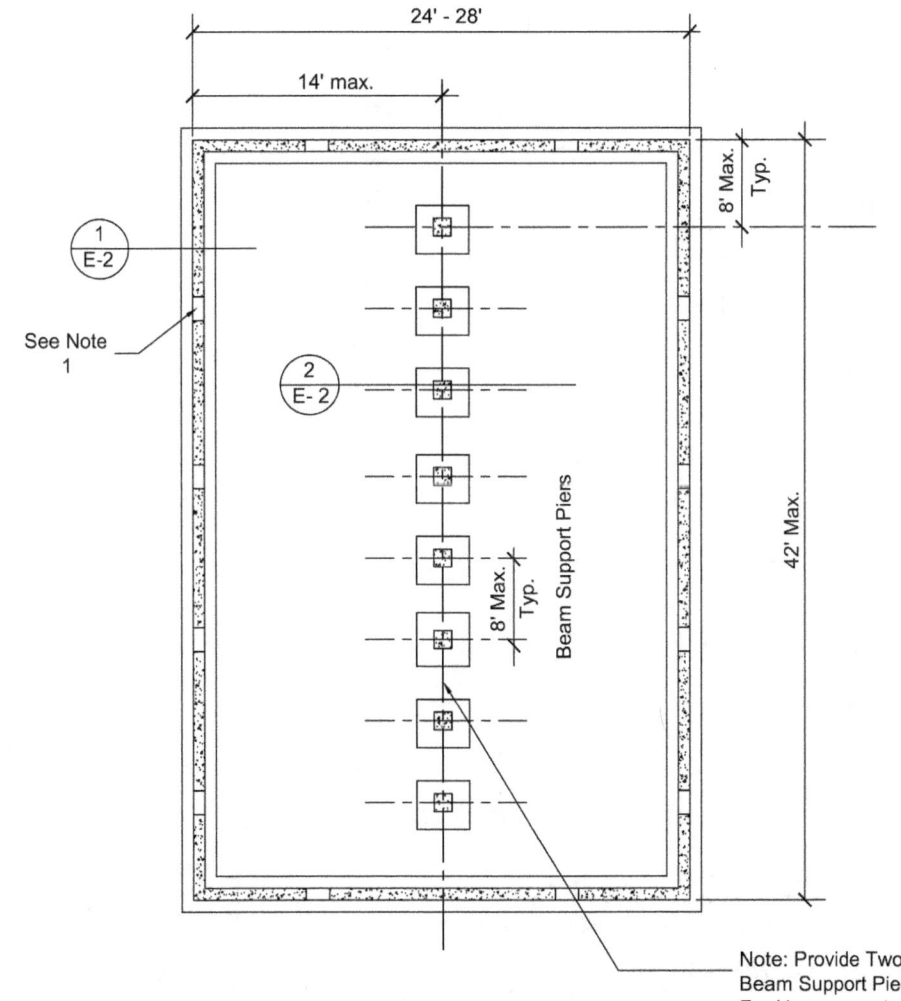

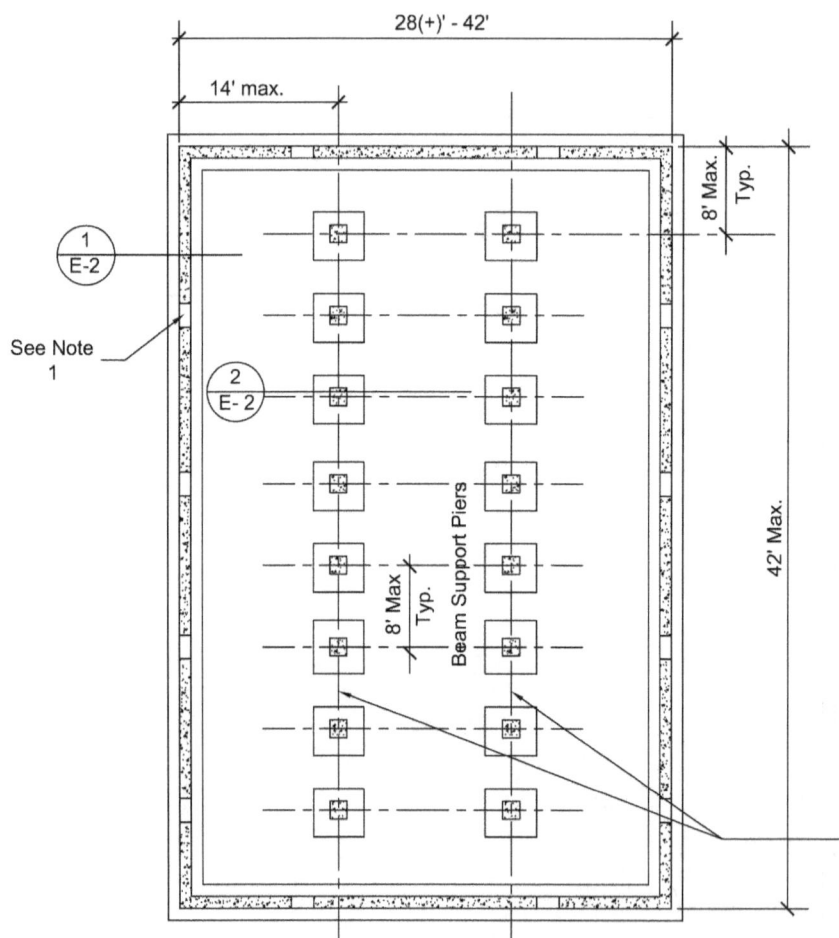

Case E - Closed Foundation Reinforced Masonry Foundation - Crawlspace
Typical Module Plan (2/E-1)
SCALE: NTS
See Sheet GN-1 for General Notes

Note 1:
1) Total area of all Flood Relief Openings shall be a minimum of 1 square inch per square foot of floor area.
2) Bottom of Flood Relief Opening shall be located no more than 1' above adjacent grade.
3) Flood Relief Opening shall comply with NFIP Standards.

Note: Provide Two Rows of Beam Support Piers For Homes greater than 28' deep

Case E - Closed Foundation Reinforced Masonry Foundation - Crawlspace	
DRAWING NO.: E-1	SHEET 27 OF 31
DATE: August 8, 2006	
REVISED:	REV. NO.

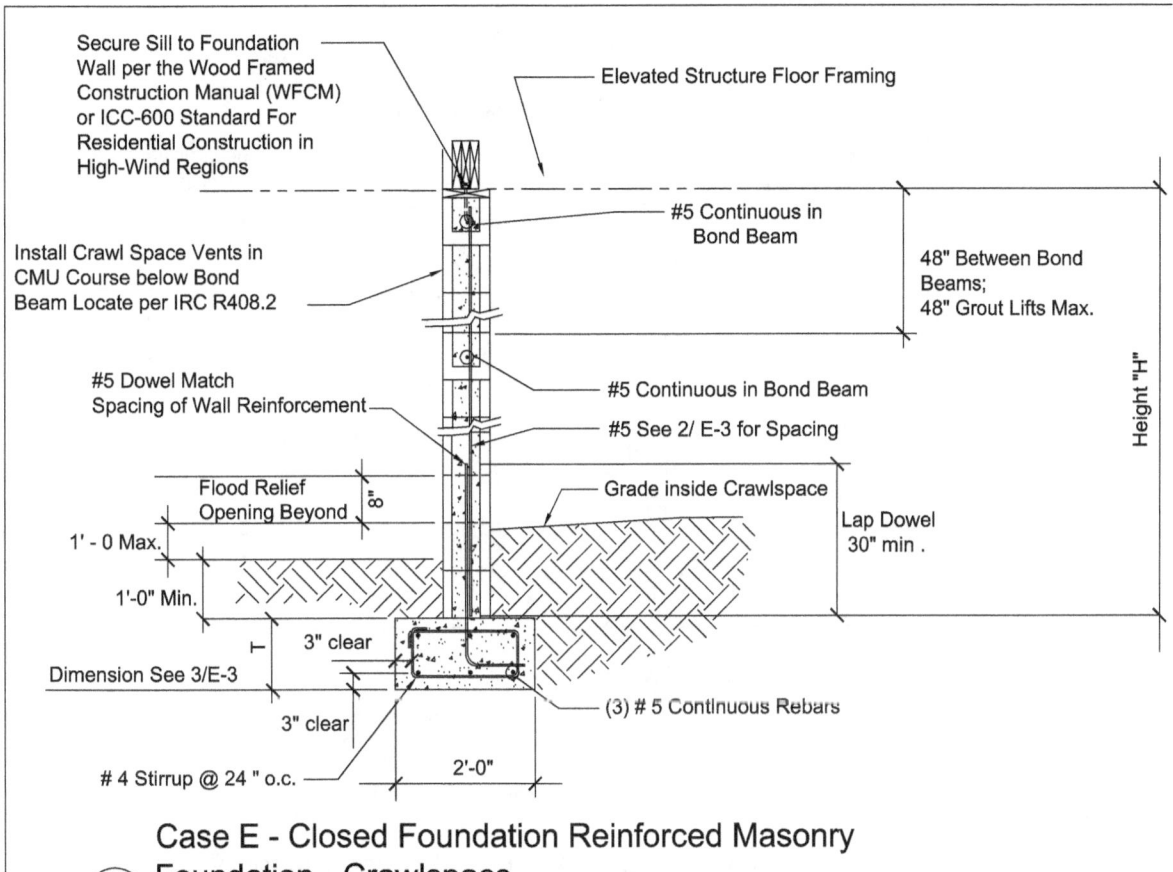

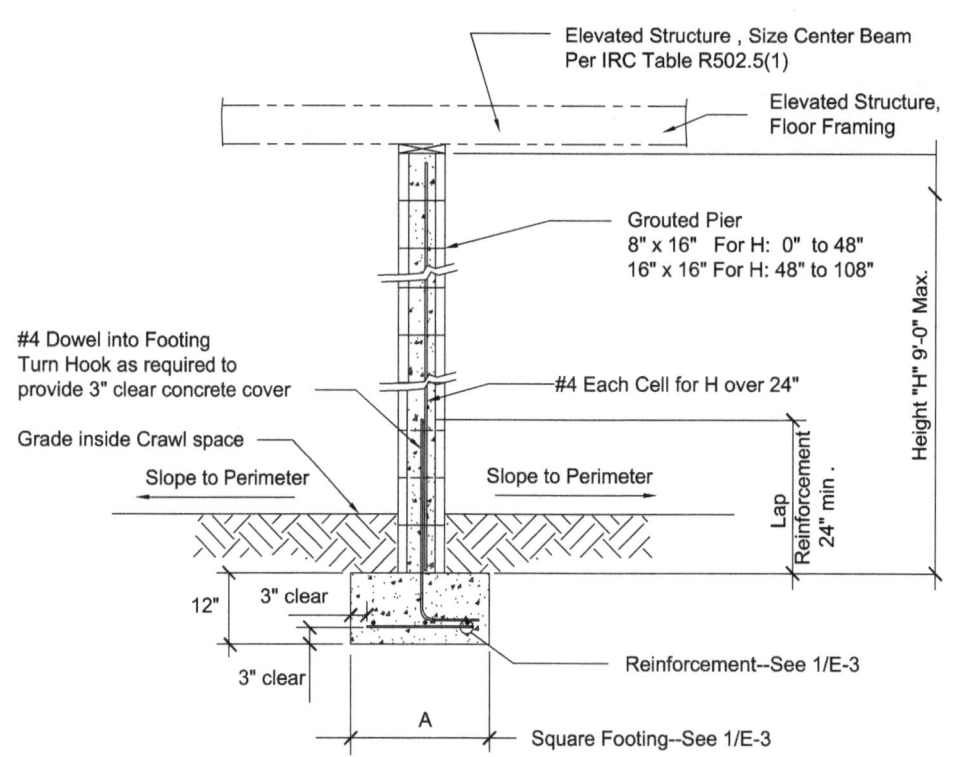

Case E - Closed Foundation Reinforced Masonry Foundation - Crawlspace
(2/E-2) **Typical Pier & Footing Section**
SCALE: NTS

Case E - Closed Foundation Reinforced Masonry Foundation - Crawlspace	
DRAWING NO.: E-2	SHEET 28 OF 31
DATE: August 8, 2006	
REVISED:	REV. 9

FEMA

Note 1:
1) Total area of all Flood Relief Openings shall be a minimum of 1 square inch per square foot of floor area.
2) Bottom of Flood Relief Opening shall be located no more than 1'-0" above adjacent grade.
3) Flood Relief Opening shall comply with NFIP Standards.

Column Spacing	Single Story Footing Size - A	Single Story Reinf. Size	Two Story Footing Size - A	Two Story Reinf. Size
4'-0"	24"	(3) # 4	30"	(3) # 4
5'-0"	26"	(3) # 4	32"	(3) # 4
6'-0"	28"	(3) # 4	34"	(3) # 4
7'-0"	28"	(3) # 5	36"	(3) # 5
8'-0"	30"	(3) # 5	38"	(3) # 5

Case E - Closed Foundation Reinforced Masonry Foundation - Crawlspace
(1/E-3) **Size and Reinforcement Schedule**

Wall Height	Wind Speed (mph) 150	140	130	120
0'-8"	40"	32"	24"	18"
1'-4"	38"	30"	24"	16"
2'-0"	36"	28"	22"	16"
2'-8"	34"	26"	22"	14"
3'-4"	34"	26"	22"	12"
4'-0"	32"	24"	20"	12"
6'-0"	28"	20"	18"	12"
8'-0"	24"	12"	14"	12"

Notes:
1) Grout all cells in foundations for homes placed in 140 mph and 150 mph wind zones and all cells in foundations that are 4 feet tall or greater.
2) In 120 mph and 130 mph wind zones, grout all cells containing rebar in crawl space foundations

Case E - Closed Foundation Reinforced Masonry Foundation Crawlspace
(3/E-3) **Table 2 Perimeter Footing Thickness (T) Required to Resist Uplift**

	Single Story		Two Story	
Wall Height	8" CMU	12" CMU	8" CMU	12" CMU
2'-0"	72"	72"	56"	72"
4'-0"	56"	72"	48"	56"
6'-0"	40"	48"	32"	40"
8'-0"	24"	32"	24"	32"

Case E - Closed Foundation Reinforced Masonry Foundation - Crawlspace

(2/E-3) Spacing (Vertical Bars) Crawl Space Foundation Wall #5 bars

Case E - Closed Foundation Reinforced Masonry Foundation - Crawlspace	
DRAWING NO.: E -3	SHEET 29 OF 31
DATE: August 8, 2006	
REVISED:	REV. 9

RECOMMENDED RESIDENTIAL CONSTRUCTION
FOR COASTAL AREAS

Building on Strong and Safe Foundations

B. Pattern Book Design

The illustrations in this appendix are from *A Pattern Book for Gulf Coast Neighborhoods* prepared for the Mississippi Governor's Rebuilding Commission on Recovery, Rebuilding, and Renewal by Urban Design Associates (UDA) of Pittsburgh, Pennsylvania, in November 2005.

http://www.mississippirenewal.com/documents/Rep_PatternBook.pdf

PATTERN BOOK DESIGN **B**

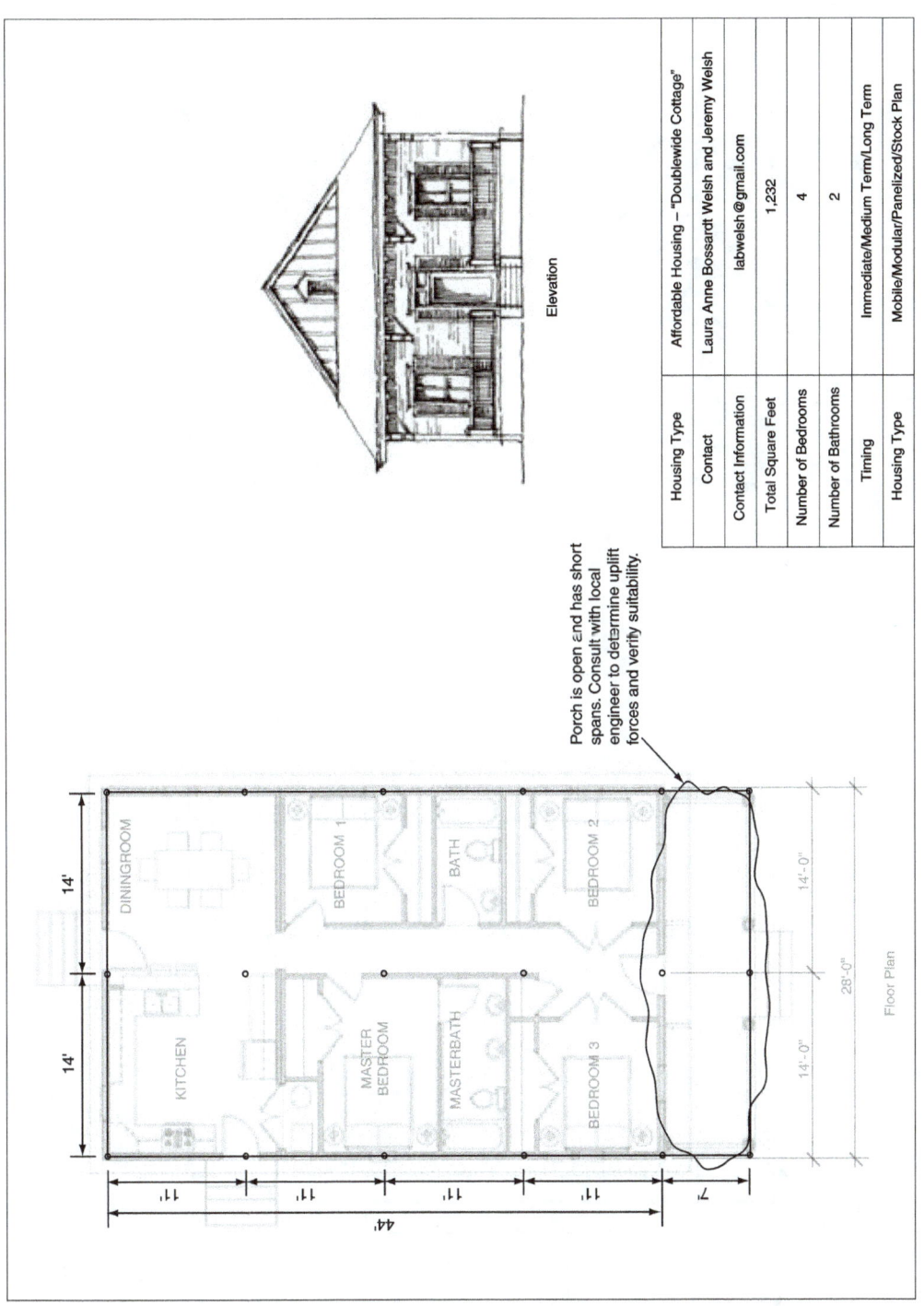

Figure B-1.
Pattern Book for Gulf Coast Neighborhoods
Page 63 Top

B PATTERN BOOK DESIGN

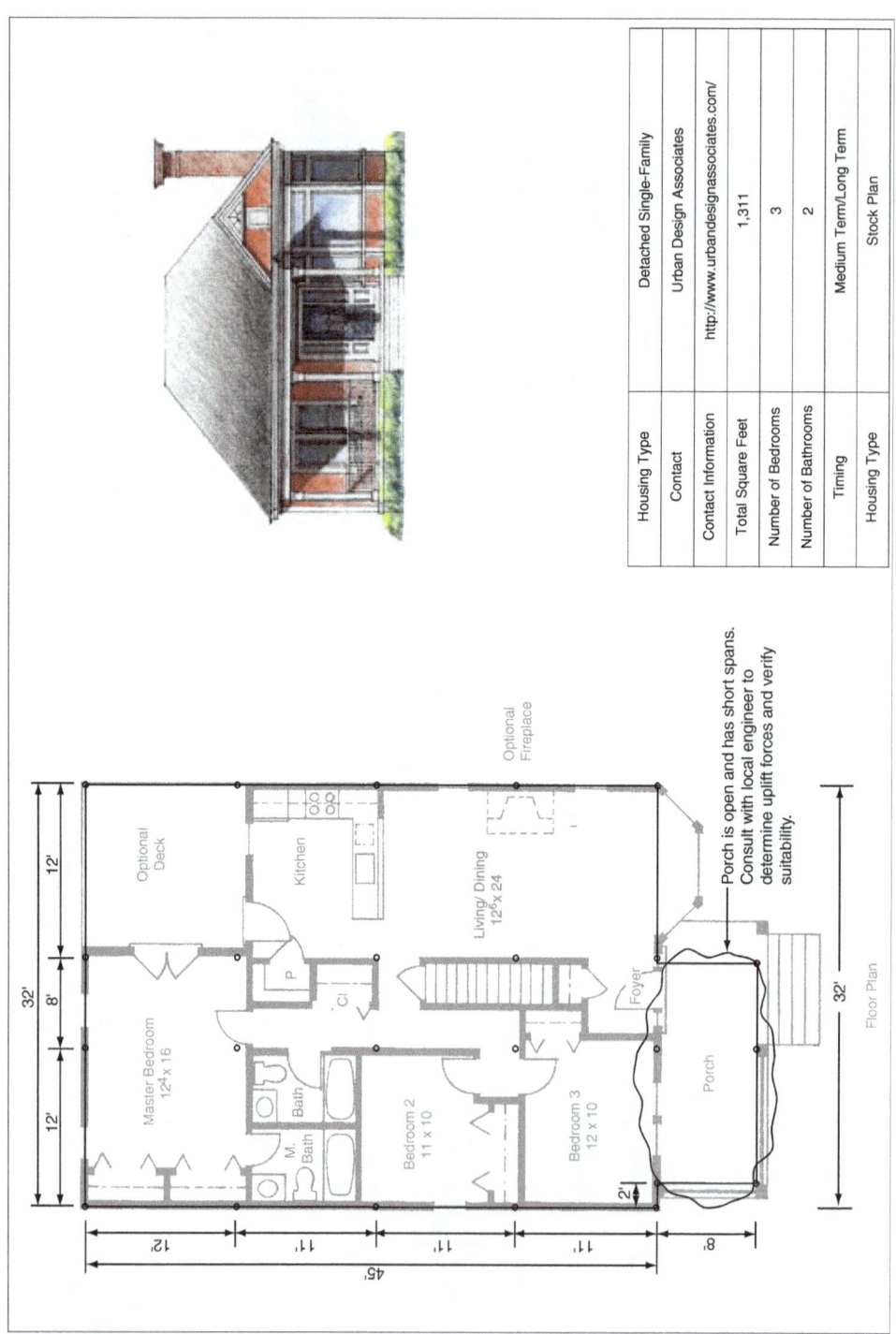

Figure B-2.
Pattern Book for Gulf Coast Neighborhoods
Page 63 Bottom

PATTERN BOOK DESIGN **B**

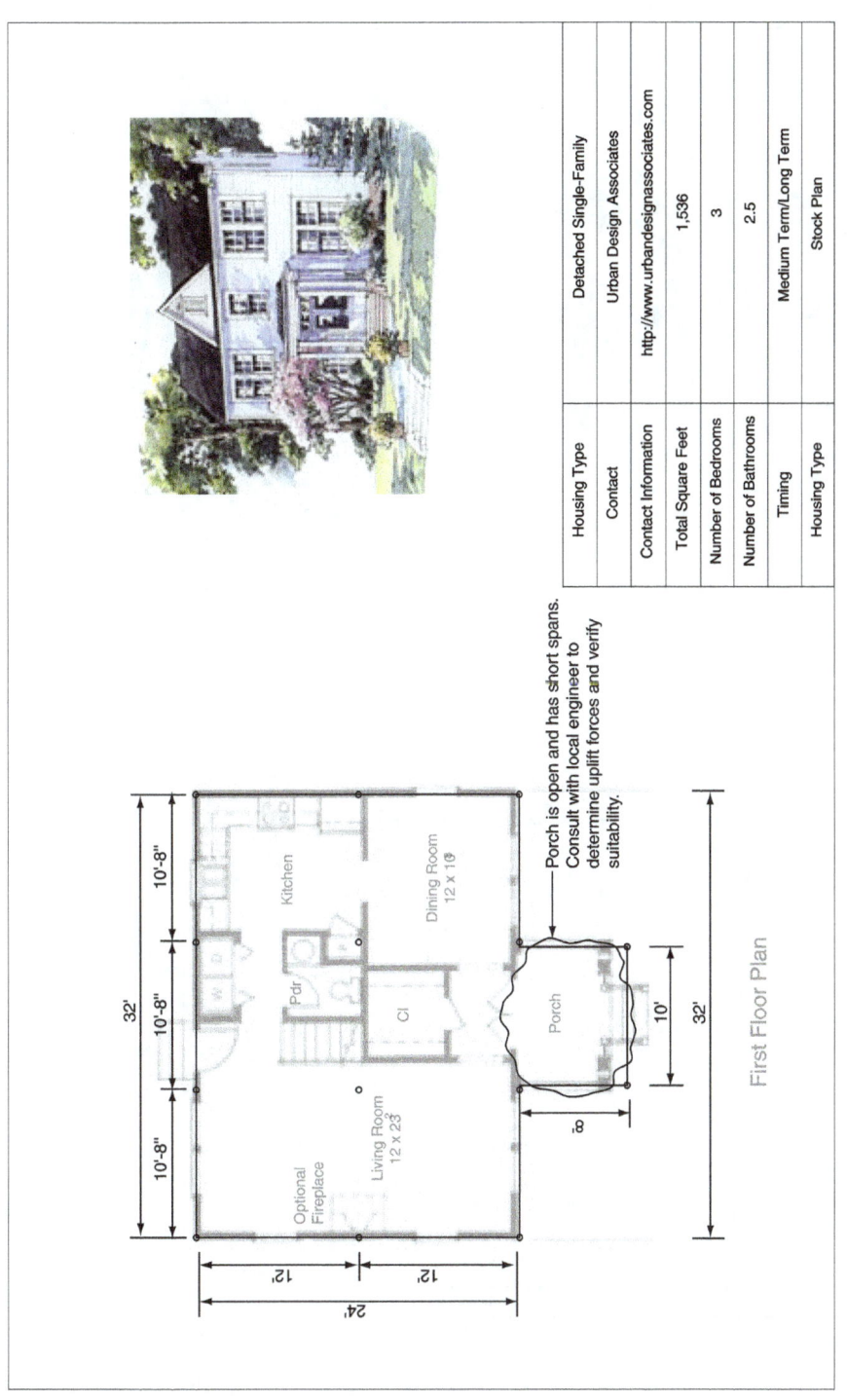

Figure B-3.
Pattern Book for Gulf Coast Neighborhoods
Page 64 Bottom

B PATTERN BOOK DESIGN

Figure B-4.
Pattern Book for Gulf Coast Neighborhoods
Page 65 Bottom

PATTERN BOOK DESIGN B

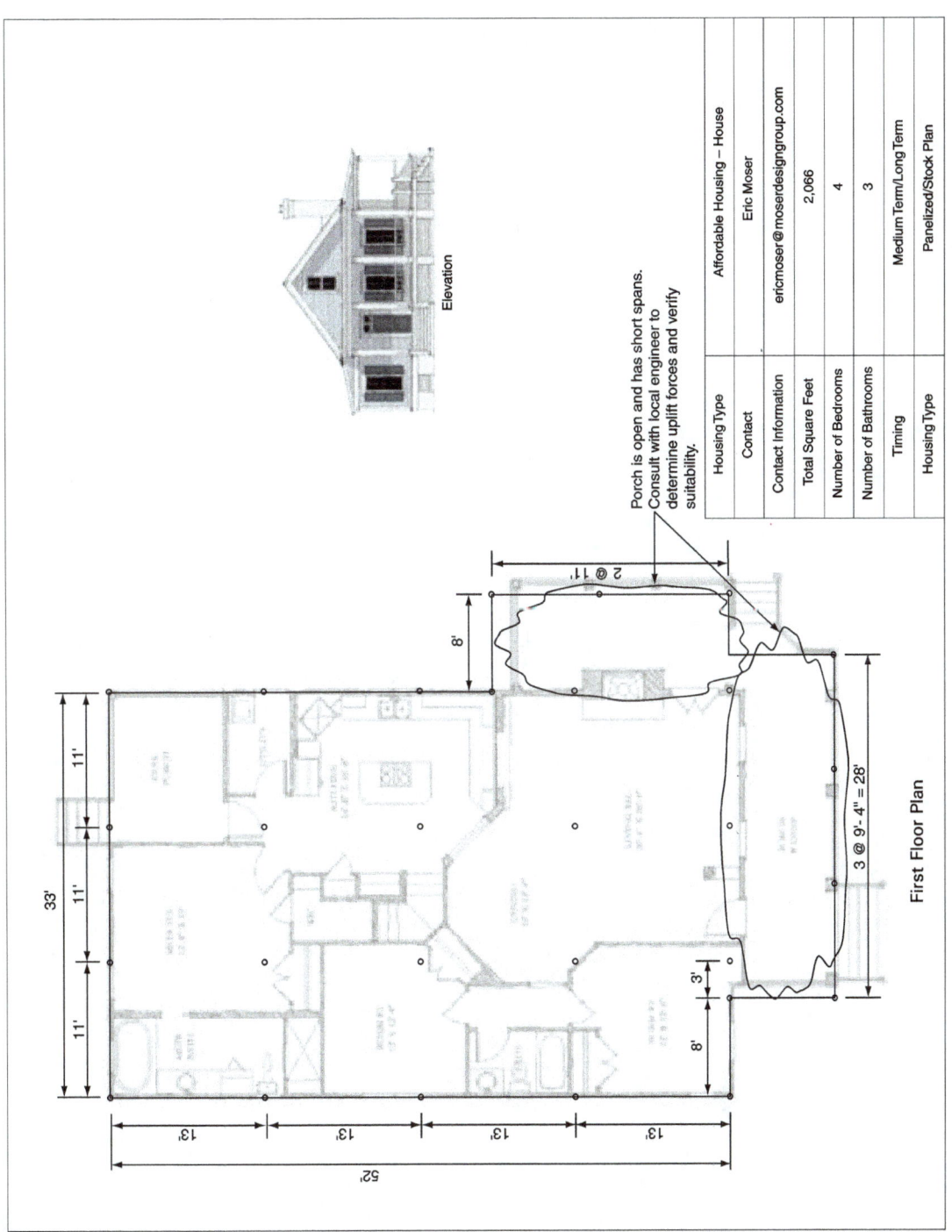

Figure B-5.
Pattern Book for Gulf Coast Neighborhoods
Page 67 Top

RECOMMENDED RESIDENTIAL CONSTRUCTION
FOR COASTAL AREAS

Building on Strong and Safe Foundations

C. Assumptions Used in Design

Coastal Areas Foundation Designs

The foundation designs proposed in Appendix A are based on the following standards and codes:

ASCE 7-05

Minimum Design Loads for Buildings and Other Structures
American Society of Civil Engineers (ASCE)

ASCE 24

Flood Resistant Design and Construction
American Society of Civil Engineers (ASCE)

ACI 318-02

Building Code Requirements for Structural Concrete
American Concrete Institute (ACI)

C ASSUMPTIONS USED IN DESIGN

ACI 530-02/ASCE 5-02/TMS 402-02

Building Code Requirements for Masonry Structures
American Concrete Institute (ACI)
American Society of Civil Engineers (ASCE)
The Masonry Society (TMS)

ANSI/AF&PA NDS-2005

National Design Specifications for Wood Construction
American Forest & Paper Association (AF&PA)
American National Standards Institute
American National Standards Institute (ANSI)
American Wood Council (AWC)

IRC-2003

2003 International Residential Code for One- and Two-Family Dwellings
International Code Council (ICC)

FEMA 550 has been checked and found to be consistent with both the 2006 and the 2009 IRC.

To provide flexibility for the builder, a range of dead loads and building dimensions was used for calculating reactions on the foundation elements. For uplift and overturning analyses, the structure was assumed to be relatively light and narrow, and constructed with a relatively low-sloped roof. For sliding analyses, the home was considered relatively deep and constructed with a steeper roof slope. For the gravity loading analysis, a heavier structure was assumed.

Dead Loads

For Use in ASCE 7-05 ASD Uplift/Overturning Load Combination #7 (0.6D + W +H)

First Floor	8 psf	Vinyl flooring, 5/8-inch plywood sub-floor and 2 by 8 joists 16 inches on centers
Second Floor	10 psf	First floor components plus 1 layer of ½-inch gypsum drywall
Wall	9 psf	Wood siding, 2 by 4 studs 16 inches on centers, ½-inch plywood wall sheathing, and one layer of ½-inch gypsum drywall
Roof	12 psf	200 lb/sq asphalt roofing, 15 lb/sq felt, ½-inch plywood decking, 2 by 4 top and bottom truss chords 24 inches on centers, ½-inch gypsum drywall ceiling finish

Dead Loads

For Use in ASCE 7-05 ASD Gravity Load Combination #2 (D + H + F + L + T)

ASSUMPTIONS USED IN DESIGN C

First Floor	16 psf	Dead loads increased 8 lb/sf to account for additional finishes like hardwood flooring (4 lb/sf), ½-inch slate (7 lb/sf), or thin set tile (5 lb/sf)
Second Floor	18 psf	
Wall	10 psf	Wall weight increased to account for cement composite siding
Roof	12 psf	
Concrete	150 psf	Normal weight concrete. Footings for continuous perimeter walls were also sized to support full height brick veneer at 40 psf.
Masonry	115 psf	Medium weight block
Grout	105 psf	
Brick Veneer	40 psf	

Wind Loads

Designs provided for 120-mph, 130-mph, 140-mph, and 150-mph zones (3-second gust wind speeds per ASCE 7-05). Wind analysis used Method 2 for buildings of all heights.

Exposure

Category C	Open terrain with scattered obstructions generally less than 30 feet in height; shorelines in hurricane-prone areas
$K_{zt} = 1$	No topographic effects (i.e., no wind speedup effects from hills, ridges, or escarpments)
$K_d = 0.85$	Wind directionality factor (for use with ASCE 7-02 load combinations)
K_z, K_h	Velocity pressure coefficients for Exposure Category C

Flood Loads

V zone	Breaking wave load from a wave with height 78 percent of still-water depth (d_s) Flood velocity (fps) equal to $(gd_s)^{1/2}$ up to a maximum of 10 fps (FEMA 55 Upper Bound)
Coastal A zone	Breaking wave load from 1½ foot up to a 3-foot high wave Flood velocity (fps) equal to $(gd_s)^{1/2}$ up to a maximum of 5 fps (FEMA 55 Upper Bound)

C ASSUMPTIONS USED IN DESIGN

Non-Coastal A zone | Breaking wave load up to a 1½-foot high wave
Flood velocity (fps) equal to stillwater flood depth (d_s) (in feet) (FEMA 55 Lower Bound)

Lateral Loads (on stem walls)

Lateral earth pressures from saturated soils	100 pounds per cubic foot (pcf)
Surcharge for slab weight and first floor live load	65 pounds per square foot (psf)

Live Loads

First Floor	40 psf
Second Floor	30 psf
Roof	20 psf

Soil Bearing Capacity

1,500 psf presumptive value for clay, sandy clay, silty clay, clayey silt, and sandy silt (CL, ML, MH, and CH soils) (2009 IRC)

Building Dimensions

Building Width	14 ft (per module)
Building Depth	max 42 ft min 24 ft
Shear Wall Spacing	max 42 ft
Floor Height	10 ft (floor to floor dimension)
Roof Pitch Ratio	min 3:12 Uplift and overturning calculation max 12:12 Sliding calculation
Roof Overhang	2 ft

RECOMMENDED RESIDENTIAL CONSTRUCTION
FOR COASTAL AREAS

Building on Strong and Safe Foundations

D. Foundation Analysis and Design Examples

Chapter 3 described the types of loads considered in this manual. This appendix demonstrates how these loads are calculated using a sample building and foundation. The reactions from the loads imposed on the example building are calculated, the loads on the foundation elements are determined, and the total loads are summed and applied to the foundation.

There is a noteworthy difference between the approach taken for designing the foundations included in this manual and the analyses that a designer may undertake for an individual building. The analyses used for the designs in this manual were based on the "worst case" loading scenario for a "range of building sizes and weights." This approach was used to provide

D FOUNDATION ANALYSIS AND DESIGN EXAMPLES

designers and contractors with some flexibility in selecting the home footprint and characteristics for which these pre-engineered foundations would apply. This also simplifies application of the pre-engineered foundations, reducing the number of pre-engineered foundations, and results in conservative designs.

For example, the designs presented were developed to resist uplift and overturning for a relatively light structure with a flat roof (worst case uplift and overturning) while gravity loads were based on a relatively heavy structure to simulate worst case gravity loads. Sliding forces were determined for a relatively deep home with a steep roof to simulate the largest lateral loads. The range of building weights and dimensions applicable to the designs are listed in Appendix C.

Since the designs are inherently conservative for some building dimensions and weights, a local design professional may be consulted to determine if reanalyzing to achieve a more efficient design is warranted. If a reanalysis is determined to be cost-effective, the sample calculations will aid the designer in completing that analysis.

D.1 Sample Calculations

The sample calculations have been included to show one method of determining and calculating individual loads, and calculating load combinations.

Type of Building

The sample calculation is based on a two-story home with a 28-foot by 42-foot footprint and a mean roof height of 26 feet above grade. The home is located on Little Bay in Harrison County, Mississippi, approximately 1.5 miles southwest of DeLisle (see Figure D-1). The site is located on the Harrison County Flood Recovery Map in an area with an Advisory Base Flood Elevation (ABFE) of 18 feet, located between the 1.5-foot wave contour and the 3-foot wave contour (Coastal A zone). The local grade elevation is 15 feet North American Vertical Datum (NAVD). The calculations assume short- and long-term erosion will occur and the ground elevation will drop 1 foot during the life of the structure. This places the home in a Coastal A zone with the flood elevation approximately 4 feet above the eroded exterior grade. Based on ASCE 7-05, the 3-second gust design wind speed at the site is 128 miles per hour (mph) Exposure Category C. To reduce possible damage and for greater resistance to high winds, the home is being designed for a 140-mph design wind speed.

The home has a gabled roof with a 3:12 roof pitch. The home is wood-framed, contains no brick or stone veneer, and has an asphalt shingled roof. It has a center wood beam supporting the first floor and a center load bearing wall supporting the second floor. Clear span trusses frame the roof and are designed to provide a 2-foot overhang. No vertical load path continuity is assumed to exist in the center supports, but vertical and lateral load path continuity is assumed to exist elsewhere in the structure.

FOUNDATION ANALYSIS AND DESIGN EXAMPLES   D

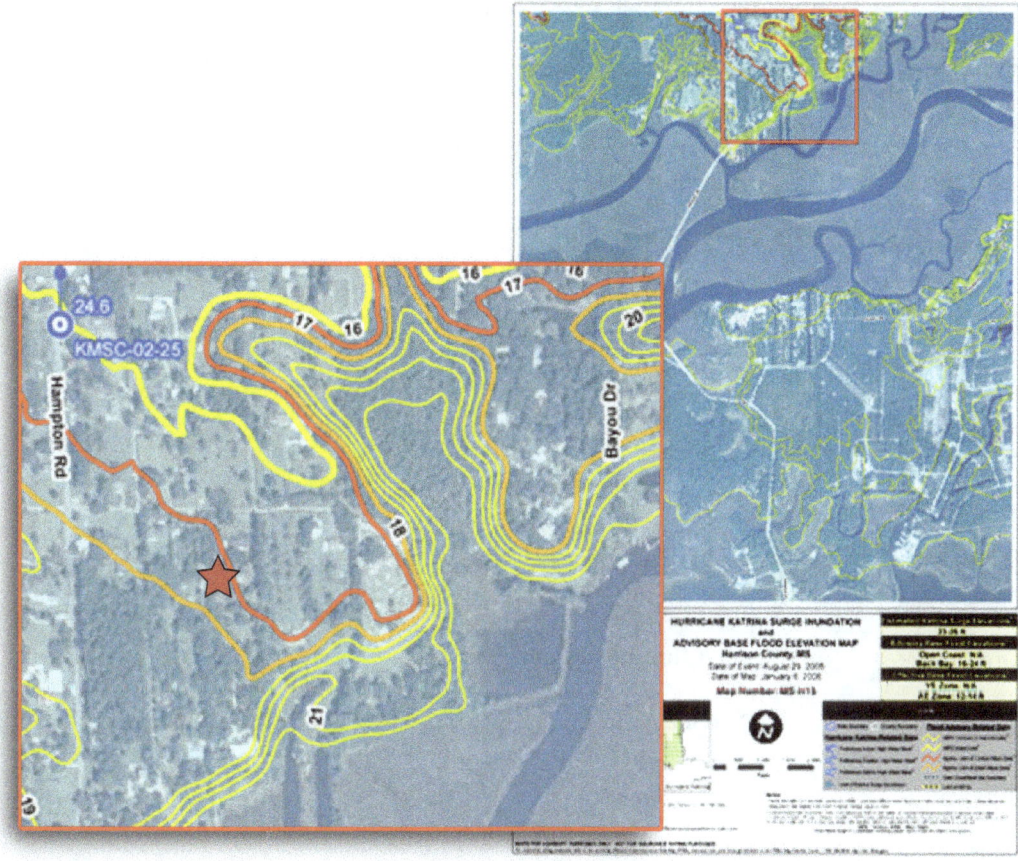

Figure D-1.
Star indicates the location of the sample calculation home on Little Bay in Harrison County, Mississippi, approximately 1.5 miles southwest of DeLisle. Inset on the left (from the map) is enlarged.

The proposed foundation for the home is a system of steel pipe piles, a reinforced concrete grade beam, and concrete columns extending from the grade beam to the elevated structure.

Methodology

1. Determine the loads based on the building's parameters (Section D.1.1)

2. Calculate wind and flood loads using ASCE 7-05 (Section D.1.2)

3. Consider the structure as a rigid body, and use force and moment equilibrium equations to determine reactions at the perimeter foundation elements (Section D.2)

D FOUNDATION ANALYSIS AND DESIGN EXAMPLES

D.1.1 Determining Individual Loads on a Structure

Building Dimensions and Weights (pulled from the text of the example problem)

B	= 42	Building width (ft)
L	= 28	Building depth (ft)
F_1	= 10	First floor height (ft)
F_2	= 10	Second floor height (ft)
r	= 3	3:12 roof pitch
W_{ovhg}	= 2	Width of roof overhang (ft)

Dead Loads

W_{rfDL}	= 12	Roof dead load including upper level ceiling finish (in pounds per square foot [psf])
W_{1stDL}	= 8	First floor dead load (psf)
W_{2ndDL}	= 10	Second floor dead load, including first floor ceiling finish (psf)
W_{wlDL}	= 9	Exterior wall weight (psf of wall area)

Live Loads

W_{1stLL}	= 40	First floor live load (psf)
W_{2ndLL}	= 30	Second floor live load (psf)
W_{rfLL}	= 20	Roof live load (psf)

Wind Loads

Building Geometry

h/L	= 1	
L/B	= 0.7	

Site Parameters (ASCE 7-05, Chapter 6)

K_{zt} = 1   Topographic factor (no isolated hills, ridges, or escarpments)

K_d = 0.85   Directionality factor (for use with ASCE 7-05, Chapter 2, Load Combinations)

K_h = 0.94   For simplicity, Velocity Pressure Coefficient used at the 26 foot mean roof height was applied at all building surfaces. See ASCE 7-05, Table 6-3, Cases 1 and 2.

I = 1   Importance factor (residential occupancy)

V = 140   3-second gust design wind speed (mph)

G = 0.85   Gust effect factor (rigid structure, ASCE 7-05, Section 6.5.8.1)

External Pressure Coefficients (C_p) (ASCE 7-05, Figure 6-6)

C_{pwwrf} = -1.0   Windward roof (side facing the wind)

C_{plwrf} = -0.6   Leeward roof

C_{pwwl} = 0.8   Windward wall

C_{plwwl} = -0.5   Leeward wall (side away, or sheltered, from the wind)

C_{peave} = -0.8   Windward eave

Positive coefficients indicate pressures acting on the surface. Negative coefficients indicate pressures acting away from the surface.

Velocity Pressure (q_h) (ASCE 7-05, Section 6.5.10)

Velocity pressure at mean roof height:

$q_h = 0.00256\, K_h K_{zt}\, K_d\, V^2\, I$

$ = 0.00256\, (0.94)(1)(0.85)(140)^2(1)$

$q_h = 40\text{ psf}$

Wind Pressures (P)

Determine external pressure coefficients for the various building surfaces. Internal pressures, which act on all internal surfaces, do not contribute to the foundation reactions. For sign convention, positive pressures act inward on a building surface and negative pressures act outward.

D FOUNDATION ANALYSIS AND DESIGN EXAMPLES

P_{wwrfV} = $q_h \, GC_{pwwrf}$ Windward roof
         = $(40)(0.85)(-1)$
P_{wwrfV} = -34 psf

Likewise

P_{lwrfV} = $q_h \, GC_{plwrf}$ Leeward roof

P_{lwrfV} = -20 psf

P_{wwwl} = $q_h \, GC_{pwwwl}$ Windward wall

P_{wwwl} = 27 psf

P_{lwwl} = $q_h \, GC_{plwwl}$ Leeward wall

P_{lwwl} = -17 psf

P_{wweave} = $q_h \, GC_{peave}$ Windward roof overhang

P_{wweave} = -27 psf   - eave
           = -34 psf   - upper surface
           = -61 psf   - total

Wind Forces (F) (on a 1-foot wide section of the home)

F_{wwrfV} = $P_{wwrfV} \, L/2$ Windward roof vertical force
          = $(-34 \text{ psf})(14 \text{ sf/lf})$
          = -476 lb/lf

F_{lwrfV} = $P_{lwrfV} \, L/2$ Leeward roof vertical force
          = $(-20 \text{ psf})(14 \text{ sf/lf})$
          = -280 lb/lf

F_{wwrfH} = $P_{wwrfH} \, L/2 \, (r/12)$ Windward roof horizontal force
          = $(-34 \text{ psf})(14 \text{ sf/lf})(3/12)$
          = -119 lb/lf

F_{lwrfH} = $P_{lwrfH} \, L/2 \, (r/12)$ Leeward roof horizontal force
          = $(-20 \text{ psf})(14 \text{ sf/lf})(3/12)$
          = -70 lb/lf

$F_{wwwl1st}$ = P_{wwwl} (F_1)   Windward wall on first floor
= (27 psf)(10 sf/lf)
= 270 lb/lf

$F_{wwwl2nd}$ = P_{wwwl} (F_2)   Windward wall on second floor
= (27 psf)(10 sf/lf)
= 270 lb/lf

$F_{lwwl1st}$ = P_{lwwl} (F_1)   Leeward wall on first floor
= (-17 psf)(10 sf/lf)
= -170 lb/lf

$F_{lwwl2nd}$ = P_{lwwl} (F_2)   Leeward wall on second floor
= (-17 psf)(10 sf/lf)
= -170 lb/lf

F_{wweave} = $P_{wweave} W_{owhg}$ Eave vertical force (lb)
= -(61 psf)(2 sf/lf) *horizontal projected areas is negligible so horizontal force is neglected*
= -122 lb/lf

D.1.2 Calculating Reactions from Wind, and Live and Dead Loads

Sum overturning moments (M_{wind}) about the leeward corner of the home. For sign convention, consider overturning moments as negative. Since vertical load path continuity is assumed not to be present above the center support, the center support provides no resistance to overturning (see Figure D-2).

M_{wind} = (-476 lb/lf)(21 ft) + (-280 lb/lf)(7 ft) + (119 lb/lf)(21.75 ft) +
(-70 lb/lf)(21.75 ft) + (-270 lb/lf)(5 ft) + (-270 lb/lf)(15 ft) +
(-170 lb/lf)(5 ft) + (-170 lb/lf)(15 ft) + (-122 lb/lf)(29 ft)

= -23,288 ft-lb/lf

Solving for the windward reaction, therefore:

W_{wind} = M_{wind} ÷ L
W_{wind} = -23,228 ft (lb/lf ÷ 28 ft)
W_{wind} = -830 lb/lf

D FOUNDATION ANALYSIS AND DESIGN EXAMPLES

The leeward reaction is calculated by either summing vertical loads or by summing moments about the windward foundation wall. Leeward reaction equals -48 lb/lf.

Lateral Wind Loads

Sum horizontal loads (Flat) on the elevated structure. (Forces to the left are positive. See Figure D-2.)

$$F_{lat} = (-119 \text{ lb/lf}) + (70 \text{ lb/lf}) + (270 \text{ lb/lf}) + (270 \text{ lb/lf}) + (170 \text{ lb/lf}) + (170 \text{ lb/lf})$$
$$= 831 \text{ lb/lf}$$

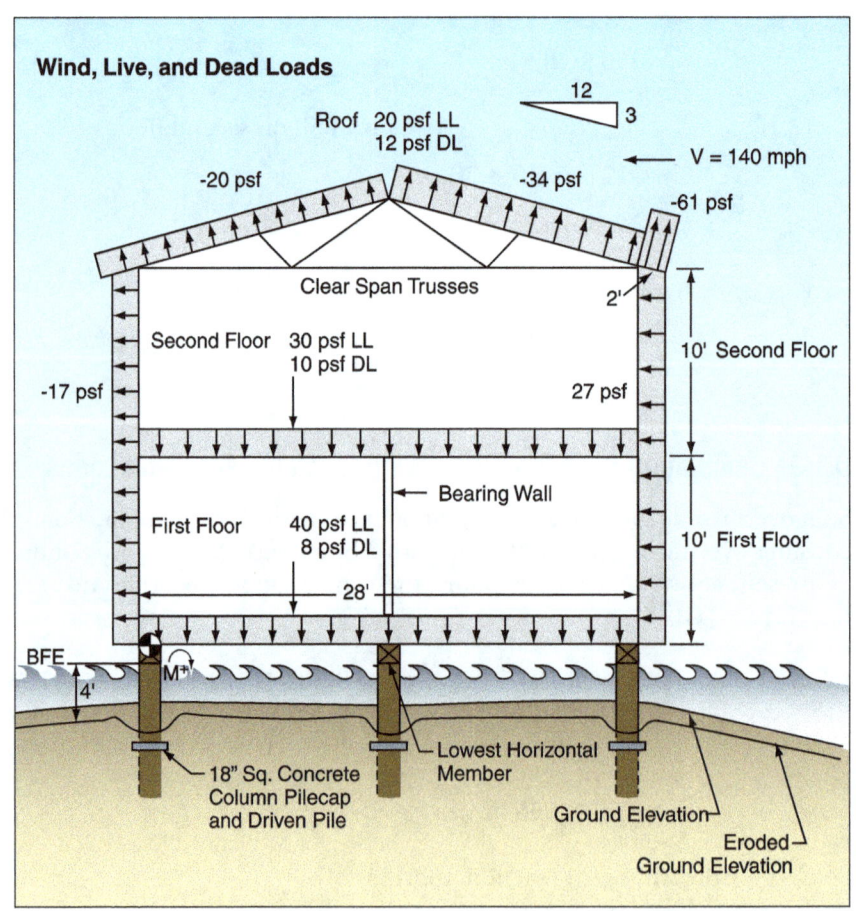

Figure D-2. Paths for wind, live, and dead loads.

Dead Loads

In this example, dead load reactions (W_{dead}) are determined by summing loads over the tributary areas. Since the roof is framed with clear-span trusses and there is a center support in the home, each exterior foundation wall supports ½ of the roof load, all of the exterior wall load, and ¼ of the first and second floor loads. This approach to analysis is somewhat conservative since it does not consider the entire dead load of the structure to resist overturning. Standard engineering practice often considers the entire weight of the structure (i.e., not just the portion supported by the perimeter foundation walls) available to resist overturning. The closed foundations in this guidance were developed considering only the tributary dead load to resist uplift. The open foundations were developed considering all dead loads to resist uplift.

$$W_{dead} = L/2\ (w_{rfDL}) + L/4\ (w_{1stDL} + w_{2ndDL}) + (F_1 + F_2)\ w_{wlDL}$$

$$= [14\ sf/lf\ (12\ psf)] + [7\ sf/lf\ (8\ psf + 10\ psf)] + [(10\ sf/lf + 10\ sf/lf)\ 9\ psf]$$

$$= 474\ lb/lf$$

Live Loads

Floor

Live loads (W_{live}) are calculated in a similar fashion

$$W_{live} = L/4(W_{1stLL} + W_{2ndLL})$$

$$= (7\ sq\ ft/lf)\ (40 + 30)\ psf$$

$$= 490\ lb/lf$$

Roof

$$W_{liveroof} = L/2(W_{rfLL})$$

$$= (14\ sq\ ft/lf)\ (20\ psf)$$

$$= 280\ lb/lf$$

D.2 Determining Load Combinations

Combine loads as specified in Chapter 2 of ASCE 7-05. For this example, an allowable stress design approach was used. A strength-based design is equally valid.

Other loads (such as snow, ice, and seismic) are listed in the ASCE 7-05 Load Combinations, but were considered to be too rare in the Gulf Coast of the United States to be considered in the design. ASCE 7-05 also lists rain loads that are appropriate for the Gulf Coast region. Since a minimum roof slope ratio of 3:12 was assumed for the homes, rain loading was not considered. Table D-1 lists foundation wall reactions for each load case. Critical reactions that contain the foundation design are highlighted.

D FOUNDATION ANALYSIS AND DESIGN EXAMPLES

Table D-1. Design Reactions on Base of Elevated Home

ASCE 7-05 Load Combination	Vertical (lb/lf)	Horizontal (lb/lf)
#1 D	474	--
#2 D + L	964	--
#3 D + L_r	754	--
#4 D + 0.75(L) + 0.75(L_r)	{1,052}	--
#5 D + W	-356	831
#6 D + 0.75(W) + 0.75(L) + 0.75(L_r)	1,016	623
#7 0.6D + W	{-546}	{831}
#8 0.6D	284	--

Where

    D   = dead load

    L   = live load

    L_r = roof live load

    W   = wind load

Note: Critical loads are in bold with brackets ({ }).

Flood Load Effects on a Foundation

In this example, since the foundation selected is a system of steel pipe piles, the equations used to calculate flood loads are based on open foundations. Some of the equations used to calculate flood loads will be different if the building has a closed foundation system.

Many flood calculations depend on the stillwater flooding depth (d_s). While not listed on FIRMs, d_s can be calculated from the BFE by knowing that the breaking wave height (H_b) equals 78 percent of the stillwater depth and that 70 percent of the breaking wave exists above the stillwater depth (see Figure 11-3 of FEMA 55). Stated algebraically:

$$\text{BFE} = \text{GS} + d_s + 0.70\, H_b$$
$$= \text{GS} + d_s + 0.70(0.78\, d_s)$$
$$= \text{GS} + 1.55\, d_s$$

$$\text{GS} = 15 \text{ ft NAVD (initial elevation)} - 1 \text{ ft (short- and long- term erosion)}$$
$$= 14 \text{ ft NAVD}$$

$$d_s = (\text{BFE} - \text{GS}) \div 1.55$$
$$= (18 \text{ feet NAVD} - 14 \text{ feet NAVD}) \div 1.55$$

= 4 ft ÷ 1.55

= 2.6 ft

Hydrostatic Loads

Hydrostatic loads act laterally and vertically on all submerged foundation elements. On open foundations, lateral hydrostatic loads cancel and do not need to be considered but vertical hydrodynamic forces (buoyancy) remain. The buoyancy forces reduce the effective weight of the foundation by the weight of the displaced water and must be considered in uplift calculations. For example, normal weight concrete which typically weighs 150 lb/ft^3 only weighs 86 lb/ft^3 when submerged in saltwater (slightly more in freshwater).

In this example, calculations are based on an 18-inch square normal weight concrete column that extends 4 feet above eroded ground elevation. The column weighs 1,350 pounds dry ((1.5 ft)(1.5 ft)(4 ft)(150 lb/ft^3). Under flood conditions, the column displaces 9 ft^3 of saltwater that, at 64 lb/ft^3, weighs 576 pounds so the column weighs only 774 pounds when submerged.

Hydrodynamic Loads

Flood Velocity

Since a Coastal A zone is close to the flood source, flood velocity is calculated using the ASCE 7-05 Equation C5-2:

$V = (g\, d_s)^{1/2}$ Upper bound flood velocity

Where

g = Gravitational acceleration (32.2 ft/sec^2)

d_s = Design stillwater depth (ft)

Hence

$V = [(32.2)(2.6)]^{1/2}$

= 9.15 feet per second (fps)

Hydrodynamic Forces

A modified version of ASCE 7-05 Equation C5-4 can be used to calculate the hydrodynamic force on a foundation element as

$F_{dyn} = \frac{1}{2} C_d\, \rho\, V^2\, A$

Where

F_{dyn} = Hydrodynamic force (lb) acting on the submerged element

D FOUNDATION ANALYSIS AND DESIGN EXAMPLES

C_d = 2.0   Drag coefficient (equals 2.0 for a square or rectangular column)

ρ = 2   Mass density of salt water (slugs/cubic foot)

A = 1.5 d_s Surface area of obstruction normal to flow (ft²)

For open foundation, "A" is the area of pier or column perpendicular to flood direction (calculated for an 18-inch square column).

Hence

F_{dyn} = (½) (2)(2)(9.15)2(1.5)(2.6)

= 653 lb/column

The force is assumed to act at a point $d_s/2$ above the eroded ground surface.

The formula can also be used for loads on foundation walls. The drag coefficient, however, is different. For foundation walls, C_d is a function of the ratio between foundation width and foundation height or the ratio between foundation width and stillwater depth. For a building with dimensions equal to those used in this example, C_d for a closed foundation would equal 1.25 for full submersion (42 feet by 4 feet) or 1.3 if submersed only up to the 2.6-foot stillwater depth.

Dynamic loads for submersion to the stillwater depth for a closed foundation are as follows:

F_{dyn} = ½ C_d ρ V^2 A

F_{dyn} = (½) (1.3)(2)(9.15)²(1)(2.6)

= 283 lb/lf of wall

Floodborne Debris Impacts

In this example, the loads imposed by floodborne debris were approximated using formula 11-9 contained in FEMA 55.

The *Commentary* of ASCE 7-05 contains a more sophisticated approach for determining debris impact loading, which takes into account the natural period of the impacted structure, local debris sources, upstream obstructions that can reduce the velocity of the floodborne debris, etc.

It is suggested that designers of coastal foundations review the later standards to determine if they are more appropriate to use in their particular design.

The FEMA 55 Formula 11.9 estimates debris impact loads as follows:

F_i = wV ÷ (g Δ t)

Where

    F_i = impact force (lb)

    w = weight of the floodborne debris (lb)

    V = velocity of floodborne debris (ft/sec)

    g = gravitational constant = 32.2 ft/sec^2

    Δt = impact duration (sec)

Floodborne debris velocity is assumed to equal the velocity of the moving floodwaters and acting at the stillwater level. For debris weight, FEMA 55 recommends using 1,000 pounds when no other data are available. The impact duration depends on the relative stiffness of the foundation and FEMA 55 contains suggested impact durations for wood foundations, steel foundations, and reinforced concrete foundations. For this example, the suggested impact duration of 0.1 second was used for the reinforced concrete column foundation.

    F_i = wV ÷ gΔt

    F_i = [1,000 lb (9.15 ft/sec)] ÷ [(32.2 ft/sec^2)(0.1 sec)]

    F_i = 2,842 lb

Breaking Wave Loads

When water is exposed to even moderate winds, waves can build quickly. When adequate wind speed and upstream fetch exist, floodwaters can sustain wave heights equal to 78 percent of their stillwater depths. Depending on wind speeds, maximum wave height for the stillwater depth at the site can be reached with as little as 1 to 2 miles of upwind fetch.

Breaking wave forces were calculated in this example using ASCE 7-05 formulae for wave forces on continuous foundation walls (ASCE 7-05 Equations 5-5 and 5-6) and on vertical piles and columns (ASCE 7-05 Equation 5-4).

The equation for vertical piles and columns from ASCE 7-05 is

    F_{brkp} = ½ C_{db} γ DH_b^2

Where

    F_{brkp} = Breaking wave force (acting at the stillwater level) (lb)

    C_{db} = Drag coefficient (equals 2.25 for square or rectangular piles/columns)

    γ = Specific weight of water (64 lb/ft^3 for saltwater)

    D = Pile or column diameter in ft for circular sections, or for a square pile or column, 1.4 times the width of the pile or column (ft). For this example, since the column is 18 inches square, D = (1.4)(1.5 ft) = 2.1 ft.

D FOUNDATION ANALYSIS AND DESIGN EXAMPLES

H_b = Breaking wave height $(0.78\ d_s)$ (ft) = $(0.78)(2.6)$ = 2.03 ft

Note: The critical angle of a breaking wave occurs when the wave travels in a direction perpendicular to the surface of the column. Waves traveling at an oblique angle (α) to the surface of the waves are attenuated by the factor $\sin^2\alpha$.

F_{brkp} = ½ $C_{db}\ \gamma\ DH_b^2$

= ½ $(2.25)(64\ lb/ft^3)(2.1\ ft)(2.03\ ft)^2$

= 623 lb

For closed foundations, use equations in Section 5.4.4.2 of ASCE 7-05 to calculate F_{brkp}. FEMA 55 contains the following two equations for calculating loads on closed foundations:

F_{brkw} = $1.1\ C_p\ \rho\ d_s^2 + 2.4\ \gamma\ d_s^2$ Case 1

and

F_{brkw} = $1.1\ C_p\ \rho\ d_s^2 + 1.9\ \gamma\ d_s^2$ Case 2

Where γ and d_s are the specific weights of water and design stillwater depths as before. C_p is the dynamic pressure coefficient that depends on the type of structure. C_p equals 2.8 for residential structures, 3.2 for critical and essential facilities, and 1.6 for accessory structures where there is a low probability of injury to human life. The term F_{brkw} is a distributed line load and equals the breaking wave load per foot of wall length where F_{brkw} is assumed to act at the stillwater elevation.

Case 1 of Formula 11.6 represents a condition where floodwaters are not present on the interior of the wall being designed or analyzed. Case 1 is appropriate for foundation walls that lack flood vents (see Figure D-3). The less stringent Case 2 is appropriate for walls where NFIP required flood vents are in place to equalize hydrostatic loads and reduce forces (see Figure D-4).

In non-Coastal A Zones, the maximum wave height is 1.5 feet. This corresponds to a stillwater depth (d_s) of approximately 2 feet (i.e., 1.5 foot/0.78 for a depth limited wave). For closed foundations in coastal areas with flood vents, a 1.5-foot breaking wave creates 1,280 lbs per linear foot of wall and 1,400 lbs per linear foot of wall on foundations that lack flood vents.

Wind Load on Columns

Wind loads have been calculated per ASCE 7-05, Section 6.5.13 (Wind Loads on Open Buildings and with Monoslope, Pitched, or Troughed Roofs). The velocity pressure (q_h) calculated previously was used, although this is a conservative figure based on the 26-foot mean roof height. The force coefficient (C_f) was determined from ASCE 7-05, Figure 6-21 (chimneys, tanks, rooftop equipment, and similar structures); ASCE 7-05, Figure 6-20 (walls and solid signs) could have been used as well.

From ASCE 7-05 Equation 6-28:

F_{wind} = $q_z\, G\, C_f\, A_f$

= 40 psf (0.85)(1.33*)(1.5 ft)(4 ft)

= 270 lb

*Interpolated C_f

Wind loads on the foundation elements are not considered in combination with flood loads since the elements are submerged during those events.

Flood Load Combinations

Section 11.6.12 of FEMA 55 provides guidance on combining flood loads. In Coastal A zones, FEMA 55 suggests two scenarios for combining flood loads. Case 1 involves analyzing breaking wave loads on all vertical supports and impact loading on one corner or critical support and Case 2 involves analyzing breaking wave loads applied to the front row of supports (row closest to the flood source), and hydrodynamic loads applied to all other supports and impact loads on one corner or critical support.

Depending on the relative values for dynamic and breaking waves, Case 1 often controls for designing individual piers or columns within a home. Case 2 typically controls for the design of the assemblage of piers or columns working together to support a home. Because of the magnitude of the load, it is not always practical to design for impact loads. As an alternative, structural redundancy can be provided in the elevated home to allow one pier or column to be damaged by floodborne debris impact without causing collapse or excessive deflection.

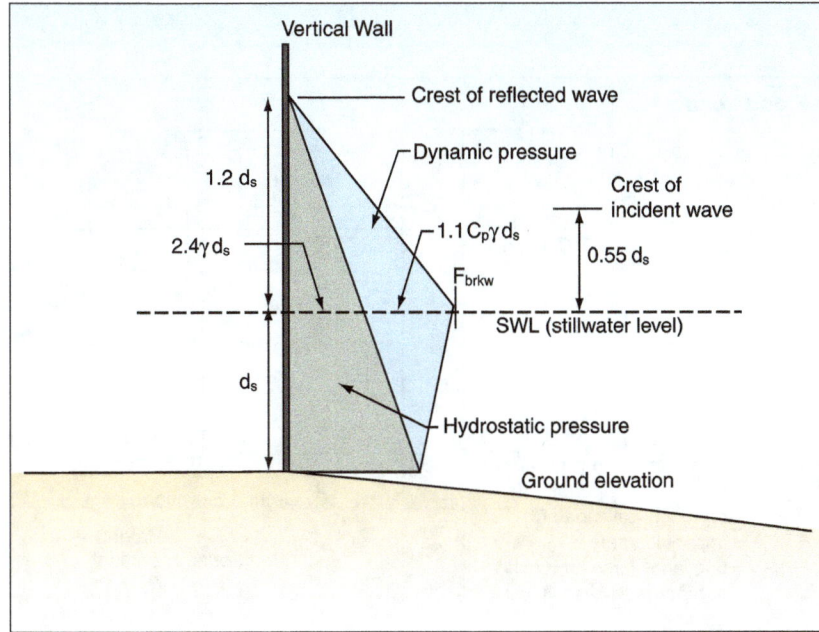

Figure D-3. Case 1. Normally incident breaking wave pressures against a vertical wall (space behind vertical wall is dry).

SOURCE: ASCE 7-05

D FOUNDATION ANALYSIS AND DESIGN EXAMPLES

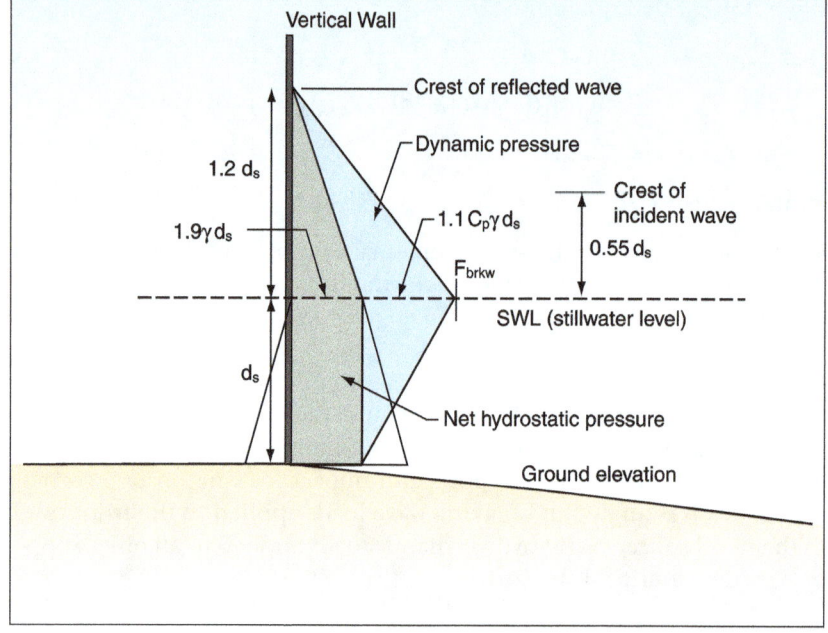

Figure D-4. Case 2. Normally incident breaking wave pressures against a vertical wall (stillwater level equal on both sides of wall).

SOURCE: ASCE 7-05

For the sample calculations, Case 1 was used (see Figure D-5) with a breaking wave load of 623 pounds applied to a non-critical column. The loads were then determined and summarized. Since the calculations must combine distributed loads on the elevated structure and discrete loads on the columns themselves, a column spacing of 7 feet is assumed in the calculations. For lateral loads on the structure, calculations are based on three rows of columns sharing lateral loads (Table D-2).

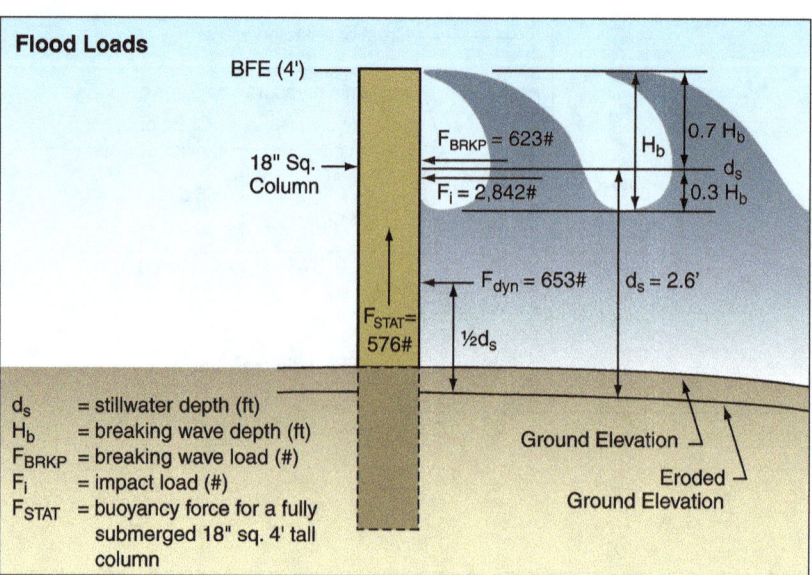

Figure D-5. Flood loads.

Table D-2. Loads on Columns Spaced 7 Feet On Center (for three rows of columns)

ASCE 7-05 Load Combination	Vertical (lb)	Horizontal (lb)
#1 D + F	1,350 + [7(474)] = 4,668	--
#2 D + F + L	1,350 + [7(964)] = 8,098	--
#3 D + F + L_r	1,350 + [7(754)] = 6,628	--
#4 D + F + 0.75(L) + 0.75(L_r)	774 + 7(1,052) = {8,138}	--
#5 D + F + W + 1.5(F_a)	774 + [7(474 - 48)] = 3,756	[(1/3)7(831)] + [1.5(623)] = {2,874}
#6 D + F + 0.75(W) + 0.75(L) + 0.75(L_r) + 1.5(F_a)	774 + 7[474 + 0.75 (-48+490+280)] = 7,883	[(1/3)7(0.75)(831)] + [1.5(623)] = 2,388
#7 0.6D + W + 1.5(F_a)	0.6[774 + 7(474)] + [7(-830)] = {-3,555}	[(1/3)7(831)] + [1.5(623)] = {2,874}
#8 0.6D	[0.6(1,350)] + [7(306)] = 3,000	--

Where

    D   = dead load

    F   = load due to fluids with well-defined pressures and maximum heights (see Section D.2 for additional information)

    F_a = flood load

    L   = live load

    L_r = roof live load

    W   = wind load

Note: Critical loads are in bold with brackets ({ }).

Results

Each perimeter column needs to support the following loads:

    Vertical Load = 8,138 lb

    Uplift   = 3,555 lb

    Lateral Load = 2,874 lb

With the critical loads determined, the foundation elements and their connections to the home can be designed.

The following two examples are to demonstrate designs using information provided in this manual. The first example is based on a closed foundation; the second example is based on an open foundation.

D FOUNDATION ANALYSIS AND DESIGN EXAMPLES

D.3 Closed Foundation Example

A structure to be supported by the closed foundation is identical to the structure analyzed in the example from Section D.1. The site, however, is different. For the closed foundation design, the structure is to be placed in a non-Coastal A zone where breaking waves are limited to 1.5 feet. The design stillwater depth is 2 feet, and the BFE is 3 feet above exterior grade. While the structure could be placed on a 3-foot foundation, the property owner requested additional protection from flooding and a 4-foot tall foundation is to be built. Since the home elevation is identical to that in the example, the loads and load combinations listed in Table D-1 are identical. However, since the foundation is closed, flood forces must first be analyzed.

Like the previous analysis example, flood forces consist of hydrodynamic loads, debris loads, and breaking wave loads. Since the home is located in a non-Coastal A zone, it is appropriate to use lower bound flood velocities. This will significantly reduce hydrodynamic and debris loads. From FEMA 55, the following equation is used:

$$V_{lower} = d_s$$
$$= 2 \text{ ft/sec}$$

Hydrodynamic Loads

$$F_{dyn} = \tfrac{1}{2} C_d \rho V^2 A$$
$$= \tfrac{1}{2} C_d \rho V^2 (1) d_s$$
$$F_{dyn} = (\tfrac{1}{2})(1.4)(2)(2)^2(1)(2)$$
$$= 11 \text{ lb/lf of wall}$$

Where C_d of 1.4 is for a (width of wall/d_s) ratio of 21 (42 ft/2 ft)

(From FEMA 55, Table 11.2)

The hydrodynamic load can be considered to act at the mid-depth of the stillwater elevation. The hydrodynamic load is less than the 27 psf wind load on the windward wall.

Debris Loads

$$F_i = wV \div gt$$
$$F_i = [1{,}000 \text{ lb } (2 \text{ ft/sec})] \div [(32.2 \text{ ft/sec}^2)(0.1 \text{ sec})]$$
$$F_i = 620 \text{ lb}$$

Due to load distribution, the impact load will be resisted by a section of the wall. Horizontal shear reinforcement will increase the width of the section of wall available to resist impact. For this example, a 3-foot section of wall is considered to be available to resist impact. The debris impact load becomes:

F_{iwall} = (1/3) 620 lb

F_{iwall} = 210 lb/ft

Breaking Wave Loads

The home is to be constructed in an SFHA; hence, the NFIP required flood vents will be installed. The breaking wave load can be calculated using formulae for equalized flood depths (Case 2).

F_{brkw} = $1.1 \, C_p \, \gamma \, d_s^2 + 1.91 \, \gamma \, d_s^2$

F_{brkw} = $\gamma \, d_s^2 \, (1.1 \, C_p + 1.91)$

F_{brkw} = $(64)(2^2)\{(1.1)(2.8) + 1.91\}$

F_{brkw} = 1,280 lb/lf

The breaking wave load can be considered to act at the 2-foot stillwater depth (d_s) above the base of the foundation wall.

The foundation must resist the loads applied to the elevated structure plus those on the foundation itself. Chapter 2 of ASCE 7-05 directs designers to include 75 percent of the flood load in load combinations 5, 6, and 7 for non-Coastal A zones. Table D-1 lists the factored loads on the elevated structure.

Critical loads from Table D-1 include 546 lb/lf uplift, 1,052 lb/lf gravity, and 831 lb/lf lateral from wind loading. The uplift load needs to be considered when designing foundation walls to resist wind and flood loads and when sizing footings to resist uplift; the gravity load must be considered when sizing footings and the lateral wind and flood loads must be considered in designing shear walls.

Extending reinforcing steel from the footings to the walls allows the designer to consider the wall as a propped cantilever fixed at its base and pinned at the top where it connects to the wood framed floor framing system. The foundation wall can also be considered simply supported (pinned at top and bottom). The analysis is somewhat simpler and provides conservative results.

The 1,280 lb/lf breaking wave load is the controlling flood load on the foundation. The probability that floodborne debris will impact the foundation simultaneously with a design breaking wave is low so concurrent wave and impact loading is not considered. Likewise, the dynamic load does not need to be considered concurrently with the breaking wave load and the 27 lb/sf wind load can not occur concurrently on a wall submerged by floodwaters.

The breaking wave load is analyzed as a point load applied at the stillwater level. When subjected to a point load (P), a propped cantilevered beam of length (L) will produce a maximum moment of 0.197 (say 0.2) PL. The maximum moments occur when "P" is applied at a distance 0.43L from the base. For the 4-foot tall wall, maximum moment results when the load is applied near the stillwater level (d_s). In this example, the ASCE 7-05 required flood load of 75 percent of the breaking wave load will create a bending moment of:

D FOUNDATION ANALYSIS AND DESIGN EXAMPLES

$$M = (0.2) f_{brkw} (L)$$
$$= (0.2)(0.75)(1{,}280 \text{ lb/ft})(4 \text{ ft})$$
$$= 768 \text{ ft-lb/lf or}$$
$$= 9{,}200 \text{ in-lb/lf}$$

The reinforced masonry wall is analyzed as a tension-compression couple with moment arm "jd," where "d" is the distance from the extreme compression fiber to the centroid of the reinforcing steel, and "j" is a factor that depends on the reinforcement ratio of the masonry wall. While placing reinforcing steel off center in the wall can increase the distance (d) (and reduce the amount of steel required), the complexity of off-center placement and the inspections required to verify proper placement make it disadvantageous to do so. For this design example, steel is considered to be placed in the center of the wall and "d" is taken as one half of the wall thickness. For initial approximation, "j" is taken as 0.85 and a nominal 8-inch wall with a thickness of 7-5/8 inches is assumed.

Solving the moment equation is as follows:

$$M = T (jd)$$
$$T = M/(jd) = \text{Tension force}$$
$$= M/(j)(t/2) \quad (t = \text{thickness of wall})$$
$$T = (9{,}200 \text{ in-lb/lf}) \div \{(0.85)(7.63 \text{ in})(0.5)\}$$
$$= 2{,}837 \text{ lb/lf}$$

For each linear foot of wall, steel must be provided to resist 2,837 pounds of bending stress and 546 pounds of uplift.

$$F_{steel} = 2{,}837 \text{ lb/lf (bending)} + 546 \text{ lb/lf (uplift)}$$
$$= 3{,}383 \text{ lb/lf}$$

The American Concrete Institute (ACI) 530 allows 60 kips per square inch (ksi) steel to be stressed to 24 ksi so the reinforcement needed to resist breaking wave loads and uplift is as follows:

$$A_{steel} = 3{,}383 \text{ lb/lf} \div 24{,}000 \text{ lb/in}^2$$
$$= 0.14 \text{ in}^2 \text{ /lf}$$

Placing #5 bars (at 0.31 in²/bar) at 24 inches on centers will provide the required reinforcement. To complete the analysis, the reinforcement ratio must be calculated to determine the actual "j" factor and the stresses in the reinforcing steel need to be checked to ensure the limits dictated in ACI 530 are not exceeded. The wall design also needs to be checked for its ability to resist the lateral forces from flood and wind.

Footing Sizing

The foundation walls and footings must be sized to prevent overturning and resist the 546 lb/lf uplift. ASCE 7-05 load combination 6 allows 60 percent of the dead load to be considered in resisting uplift. Medium weight 8-inch masonry cores grouted at 24 inches on center weigh 50 lb/sf or, for a 4-foot tall wall, 200 lb/lf. Sixty percent of the wall weight (120 lb/lf) reduces the amount of uplift the footing must resist to 426 lb/lf. At 90 lb/ft^3 (60 percent of 150 lb/ft^3 for normal weight concrete), the footing would need to have a cross-sectional area of 4.7 square feet. Grouting all cores increases the dead load to 68 lb/sf and reduces the required footing area to 4.25 square feet. The bearing capacity of the soils will control footing dimensions. Stronger soils can allow narrower footing dimensions to be constructed; weaker soils will require wider footing dimensions.

The design also needs to be checked to confirm that the footings are adequate to prevent sliding under the simultaneous action of wind and flood forces. If marginal friction resistance exists, footings can be placed deeper to benefit from passive soil pressures.

D.4 Open Foundation Example

For this example, the calculations are based on a two-story home raised 8 feet above grade with an integral slab-grade beam, mat-type foundation and a 28-foot by 42-foot footprint. The home is sited approximately 800 feet from the shore in a Coastal A zone. Subtracting the elevation of the site (determined from a topographic map or preferably from a survey) from the ABFE and adding estimated erosion (in feet) determines that the floodwaters during a design event (including wave effects and runup) will extend 6 feet above the eroded exterior grade. It is important to note, however, that submittal of an elevation certificate and construction plans to local building code and floodplain officials in many jurisdictions will require that the elevation be confirmed by a licensed surveyor referencing an established benchmark elevation.

The wood framed home has a 3:12 roof pitch with a mean roof height of 30 feet, a center wood beam supporting the first floor, and a center load bearing wall supporting the second floor. Clear span trusses frame the asphalt-shingled roof and are designed to provide a 2-foot overhang. This home is a relatively light structure that contains no brick or stone veneers.

The surrounding site is flat, gently sloping approximately 1 foot in 150 feet. The site and surrounding property have substantial vegetation, hardwood trees, concrete sidewalks, and streets. A four-lane highway and a massive concrete seawall run parallel to the beach and the established residential area where the site is located. The beach has been replenished several times in the last 50 years. Areas to the west of the site that have not been replenished have experienced beach erosion to the face of the seawall. The ASCE 7-05, 3-second gust design wind speed is 140 mph and the site is in an Exposure Category C.

The proposed foundation for the home incorporates a monolithic carport slab placed integrally with a system of grade beams along all column lines (see Figure D-6). The dimensions of the

D FOUNDATION ANALYSIS AND DESIGN EXAMPLES

grade beam were selected to provide adequate bearing support for gravity loads, resistance to overturning and sliding, and mitigate the potential of undermining of the grade beams and slab due to scour action. The home is supported by concrete columns, extending from the top of the slab to the lowest member of the elevated structure, spaced at 14 feet on center (see Figure D-7).

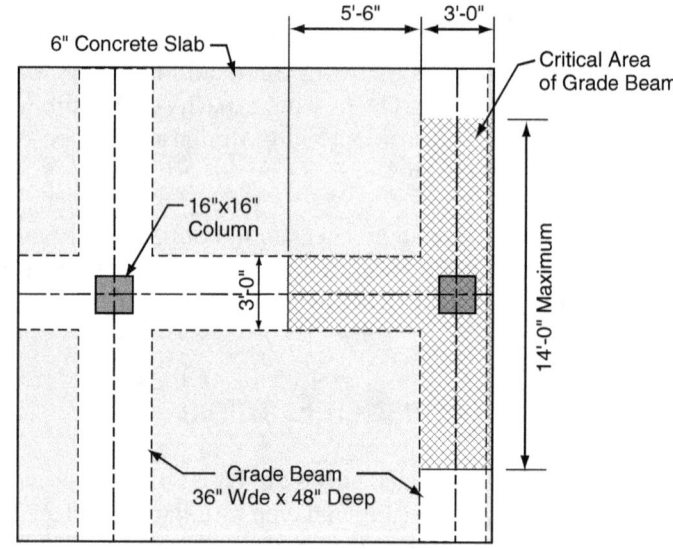

Figure D-6.
Layout of Open Foundation Example.

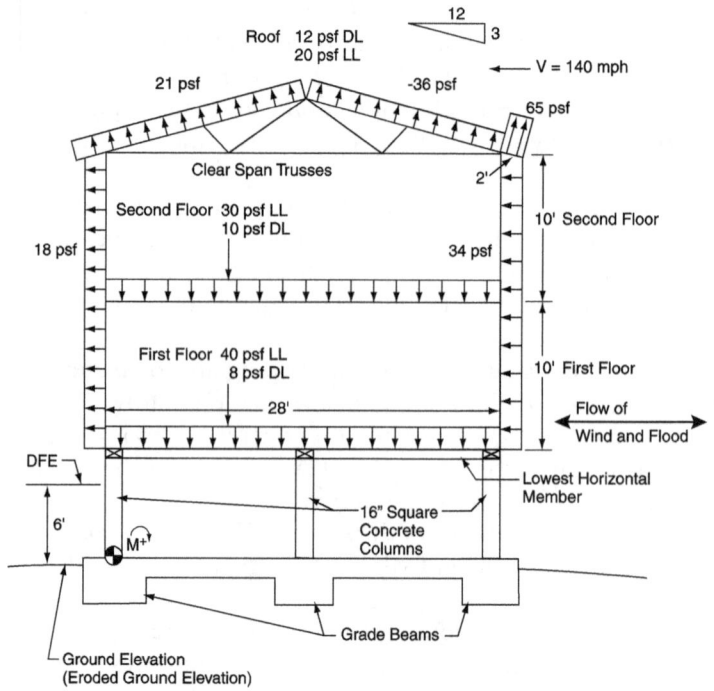

Figure D-7.
Loading Diagram for Open Foundation Example.

Lateral Wind Loads

Sum horizontal loads (F_{lat}) on the elevated structure (forces to the left are positive)

F_{lat} = (-126 lb/lf) + (74 lb/lf) + (280 lb/lf) + (280 lb/lf) + (180 lb/lf) + (180 lb/lf)

= 868 lb/lf

Dead Loads

Dead load reactions (W_{dead}) are determined by summing loads over the tributary areas. For the anterior columns:

W_{dead} = L/2 (w_{rfDL}) + L/4 ($w_{1stDL} + w_{2ndDL}$) + ($F_1 + F_2$) w_{wlDL}

= [14 sf/lf (12 psf)] + [7 sf/lf (8 psf + 10 psf)] + [(10 sf/lf + 10 sf/lf) 9 psf]

= 474 lb/lf

Live Loads

Floor

Live loads (W_{live}) are calculated in a similar fashion

W_{live} = L/4($W_{1stLL} + W_{2ndLL}$)

= (7 sq ft/lf)(40 + 30) psf

= 490 lb/lf

Roof

$W_{liveroof}$ = L/2(W_{rfLL})

= (14 sq ft/lf)(20 psf)

= 280 lb/lf

A minimum roof slope of 3:12 was assumed for the homes; rain loading was not considered.

Flood Effects

Since the foundation selected is a system of concrete columns, the equations used to calculate flood loads are based on open foundation. The stillwater flooding depth (d_s) is as follows:

d_s = DFE ÷ 1.55

= 6 ft ÷ 1.55

= 3.9 ft

Hydrostatic Loads

Calculations are based on a 16-inch square normal weight concrete column that extends 8 feet above the concrete slab.

D FOUNDATION ANALYSIS AND DESIGN EXAMPLES

The column weighs 2,123 pounds dry ((1.33 ft) (1.33 ft) (8 ft) (150 lb/ft^3)).

Under flood conditions, the column displaces 10.6 ft^3 of saltwater which, at 64 lb/ft^3, weighs 679 pounds so the column weighs 1,444 pounds when submerged.

Hydrodynamic Loads

Flood Velocity

Since a Coastal A zone is close to the flood source, flood velocity is calculated using the ASCE 7-05 Equation C5-2 as follows:

$$V = [(32.2 \text{ ft/sec}^2)(3.9 \text{ ft})]^{1/2}$$
$$= 11.21 \text{ feet per second (fps)}$$

Flood Force

ASCE 7-05 Equation C5-4 is as follows:

$$F_{dyn} = \tfrac{1}{2} C_d \rho V^2 A$$
$$= (\tfrac{1}{2})(2)(2)(11.21 \text{ fps})^2 (1.33 \text{ ft})(3.9 \text{ ft})$$
$$= 1,303 \text{ lb/column}$$

Floodborne Debris Impact

The flood debris impact can be estimated, per FEMA 55 Formula 11.9, as follows:

$$F_i = wV \div gt$$
$$= [1,000 \text{ lb }(11.21 \text{ ft/sec})] \div [(32.2 \text{ ft/sec}^2)(0.1 \text{ sec})]$$
$$= 3,478 \text{ lb}$$

Breaking Wave Load

The equation for vertical piles and columns from ASCE 7-05 is as follows:

$$F_{brkp} = \tfrac{1}{2} C_{db} \gamma D H_b^2$$
$$= \tfrac{1}{2} (2.25)(64 \text{ lb/ft}^3)(1.82 \text{ ft})(3.04 \text{ ft}^*)^2$$
$$= 1,211 \text{ lb}$$

Wind Load on Columns

For a load case combining both wind and flood forces, the column would be almost completely submerged; therefore, the wind load on the column shall not be included.

---

[*] A wave height of 3.04 ft suggests a V zone but, in this example, the depth of water is increased by erosion, which is not considered in mapping A zones. The deeper water supports a bigger wave, which in this case exceeds the V-zone wave height minimum.

Calculating Reactions from Wind, Live, and Dead Loads

Sum overturning moments (M_{wind}) and (M_{flood}) about the leeward corner of the mat foundation. For sign convention, consider overturning moments as negative. Note in this example the home is slightly higher above grade and hence the wind loads are slightly higher.

M_{wind} = (-504 lb/lf) (21 ft) + (-294 lb/lf) (7 ft) + (126 lb/lf) (21.75 ft) + (-74 lb/lf) (21.75 ft) + (-280 lb/lf) (13 ft) + (-280 lb/lf) (23 ft) + (-180 lb/lf) (13 ft) + (-180 lb/lf) (23 ft) + (-130 lb/lf) (29 ft)

= -31,841 ft-lb/lf

The vertical components of the reaction caused by the wind overturning moment is:

R_x = 31,841 lb ÷ 28 ft = +/- 1,137 lb/ft

M_{flood} = 1.5[((-1,211 lb) (3.9 ft)) + (2(-1,303 lb) (3.9 ft/2)) + ((-3,478 lb) (3.9 ft))] = 35,053 ft-lb/ft

The vertical component of the reaction caused by the flood overturning moment is:

R_x = 35,053 lb ÷ 28 ft = +/- 1,252 lb outboard columns

Load Combinations

Table D-3 summarizes loads on the open foundation example. Loads are listed for the eight load combinations and critical loads are highlighted.

Table D-3. Loads at Base of Columns Spaced 14 Feet On Center (for three rows of columns per bay)

ASCE 7-05 Load Combination	Vertical (lb)	Horizontal (lb)
#1 D + F	1,444 + 14(474) = 8,080	–
#2 D + F + L	1,444 + 14(964) = 14,940	–
#3 D + F + L_r	1,444 + 14(754) = 12,000	–
#4 D + F + 0.75(L) + 0.75(L_r)	8,080 + (0.75)[(14)(490+280)] = 16,165	–
#5 D + F + W + 1.5(F_a)	8,080 +/- 14(1,137) +/- 1,252 = 25,876; -9,716 windward; leeward	wind + (1.5)[F_{dyn} + F_i] [(14(868)(1/3)] + (1.5) [(1,303+3,478)] = 11,222
#6 D + F + 0.75(W) + 0.75(L) + 0.75(L_r) + 1.5(F_a)	8,080 +/- (0.75)(14)(1,137) +(0.75)(14)([(490+280)] +/- 1,252 = 2,348; **29,982** windward; leeward	(0.75) wind + (1.5)[F_{dyn} + F_i] [(0.75)(14)(868)(1/3)] + (1.5)[(1,303+3,478)] = 10,210
#7 0.6D + W + 1.5(F_a)	0.6 [2,123+14(474)] +/- 14(1,137) +/- (1.5)1,252 = **-12,541**; 23,051 windward; leeward	wind + (1.5)[F_{dyn} + F_i] [(14(868)(1/3)] + (1.5) [(1,303+3,478)] = **11,222**
#8 0.6D	[0.6((2,123) + 14(474))] = 5,255	–

Critical loads are in bold.

D FOUNDATION ANALYSIS AND DESIGN EXAMPLES

Where

D	= dead load
F	= fluid (buoyancy) load
L	= live load
L_r	= roof live load
W	= wind load
ww	= windward
lw	= leeward

Results

Each perimeter column needs to support the following loads:

Vertical Load	= 29,982 lb
Uplift	= 12,541 lb
Lateral Load	= 11,222 lb
Moment wind + f_{dyn}	= [(1/3)(14)(1,314)(8) + (1,303)(3.9/2)] ÷ 1,000 lb/kip
	= 51.6 ft-kip
Moment wind + f_{brkp}	= [(1/3)(14)(1,314)(8) + (1,113)(3.9)] ÷ 1,000 lb/kip
	= 96.9 ft-kip
Moment debris	= (3,478)(3.9) ÷ 1,000 lb/kip
	= 13.6 ft-kip
Moment wind + f_{dyn} + debris	= 51.6 ft-kip + 13.6 ft-kip
	= 65.1 ft-kip

The force is assumed to act at a point $d_s/2$ above the eroded ground surface. For concrete design, we use load factors per ASCE 7-05.

Ultimate Moment wind + f_{dyn}	= (48.6)(1.2)+(2.5)(2.0)
	= 63.4 ft-kip
Ultimate Moment wind + f_{brkp}	= (48.6)(1.2)+(4.3)(2.0)
	= 66.9 ft-kip
Ultimate Moment wind + f_{dyn} + debris	= (63.4)+13.6(2)
	= 90.6 ft-kip

Foundation Design

Overturning

The overturning moment due to wind with a typical bay of 14 feet wide is as follows:

M_{wind} = (-31,841 ft-lb/lf) (14 ft)

= -445,774 ft-lb

M_{Fa} (1.5) = (1.5)[(1,211 lb)(3.9 ft) + (2)(1,303 lb)(3.9 ft/2)]

= -14,707 ft-lb

M_o = -445,774 ft-lb - 14,707 ft-lb

= -460,481 ft-lb

In this example, it is assumed that the home and the foundation slab are reasonably symmetrical and uniform; therefore, it is assumed the center of gravity for the dead loads is at the center of the bay.

Dead load at perimeter columns

D_{ext} = (474 lb/ft)(14 ft)(2 columns)

= 13,272 lb

Dead load at an interior column:

D_{int} = (14 ft)(14 ft)(8 psf)

= 1,568 lb

Dead load of 3 columns: (3)(8 ft)(1.33 ft x 1.33 ft)(150)lb cubic ft

= 6,368 lb

Assume that only the grade beams are sufficiently reinforced to resist overturning (neglect weight of slab)

Dead load of the grade beams (area)(depth) (density of concrete)
[(28 ft x 3 ft) + (3)((11 ft)(3 ft)](4 ft)(150 lb cubic ft) = 109,800 lb

Summing the dead loads = 13,272 + 1,568 + 6,368 + 109,800 = 131,008 lb

Allowable dead load moment of 60 percent

M_d = (0.6)(131,008 lb)(14 ft)

= 1,100,476 ft-lb

Since M_{ot} = 460,481 ft-lb is very much less than 0.6 M_d = 1,100,476 ft-lb, the foundation can be assumed to resist overturning.

D FOUNDATION ANALYSIS AND DESIGN EXAMPLES

Sliding

The maximum total lateral load of wind and flood acting on the entire typical bay is as follows:

$$L_{wfa} = W + 1.5 F_a$$
$$= [(14 \text{ ft } (868 \text{ lb/ft}) + 1.5(7{,}203 \text{ lb})]$$
$$= 23{,}375 \text{ lb}$$

Sliding Resistance = $(\tan\Phi)N$ + Passive Resistance at Vertical Foundation Surfaces

Φ = internal angle of soil friction, assume Φ = 25 degrees

N = net normal force (building weight - uplift forces)

N = $(131{,}008) + (14 \text{ ft})[((16 \text{ ft})(-21 \text{ psf}) + (14 \text{ ft})(-36 \text{ psf}) + (2 \text{ ft})(-65 \text{ psf})]$

= 117,428 lb

Ignore passive soil pressure

Dead Load Sliding Resistance = $(\tan 25)(117{,}428 \text{ lb})$

= 54,758 lb

Since L_{wfa} = 23,371 lb is less than 60 percent of Dead Load Sliding Resistance = 54,758, the foundation can be assumed to resist sliding.

Soil Bearing Pressure

The simplified approach for this mat foundation assumes that only the grade beams carry loads to the soil; the slab between grade beams is not considered to contribute support. It is further assumed that the bearing pressure is uniform in the absence of wind and flood loading. The areas of the grade beams along the outboard column lines, in the direction of the flow of wind and flood, are considered the "critical areas" of the grade beam. The load combination table below indicates the bearing pressures for the ASCE 7-05 load combinations for the critical grade beam area. These load combinations are calculated to ensure that downward forces of the wind and flood moment couple do not overstress the soil. The factored dead load moment that resists overturning is of a magnitude such that there is no net uplift along critical grade beams. Table D-4 presents foundation bearing pressures for typical bays.

Self Weight of 1 square of foot Grade Beam = (4 ft)(150 lb/cubic ft) = 650 psf

Area of Critical Grade Beam = $[(3\text{ft})(14 \text{ ft})] + [(3 \text{ ft})(5.5 \text{ ft})]$

= 58.5 ft²

Weight of Critical Grade Beam = $(58.5 \text{ ft}^2)(4 \text{ ft})(150 \text{ lb/cubic ft})$

= 35,100 lb

Critical Column Uplift = 12,541 lb (load combination 7)

FOUNDATION ANALYSIS AND DESIGN EXAMPLES   D

Verification of Uplift Resistance = $[35{,}100 \text{ lb}(0.6)] - 12{,}541 \text{ lb}$

= 8,519 lb (positive load, no uplift)

Presumptive Allowable Bearing Pressure = 1,500 psf

Table D-4. Foundation Bearing Pressures for Typical Bays (for three rows of columns per bay)

ASCE 7-05 Load Combination	Combined Loads (lb)	Bearing Pressures (psf)
#1 D + F	8,080	8,090 ÷ 58.5 + 650 = 788
#2 D + F + L	14,940	14,950 ÷ 58.5 + 650 = 905
#3 D + F + L_r	12,000	12,000 ÷ 58.5 + 650 = 855
#4 D + F + 0.75(L) + 0.75(L_r)	16,165	16,165 ÷ 58.5 + 650 = 926
#5 D + F + W + 1.5(F_a)	28,104	28,104 ÷ 58.5 + 650 = 1,130
#6 D + F + 0.75(W) + 0.75(L) + 0.75(L_r) + 1.5(F_a)	29,982	29,982 ÷ 58.5 + 650 = **1,163**
#7 0.6D + W + 1.5(F_a)	23,051	23,051 ÷ 58.5 + 650 = 1,044
#8 0.6D	5,255	5,255 ÷ 58.5 + 650 = 740

Where

- D = dead load
- F = load due to fluids with well-defined pressures and maximum heights (See Section D.2 for additional information)
- F_a = flood load
- L = live load
- L_r = roof live load
- W = wind load

Note: The maximum bearing pressure is in bold (1,163 psf) and is less than the assumed 1,500 psf bearing pressure.

Design of Concrete members per ACI-318-02 Code and ASCE 7-05; Sections 2.3.2 and 2.3.3-1

Column Design

Verify that 16-inch x 16-inch column design is adequate.

Concrete strength = 4,000 psi

D FOUNDATION ANALYSIS AND DESIGN EXAMPLES

Reinforced with (4) #8 bars, grade 60 reinforcing, with 2½-inch clear cover

Note: 1,000 lb = 1.0 kip

Assume that the total wind load distributed through the floor uniformly to 3 columns.

Check combined axial and bending strength:

Ultimate Moment wind + f_{dyn} = (1.6) [(8 ft)((14 ft)(868 lb/ft)/(3))] + (2.0) [(3.9 ft/2)(1,303) lb)]

= 51,849 ft-lb + 5,082 ft-lb

= 56,931 ft-lb ÷ 1,000 lb/kip = 56.9 ft-kip

Ultimate Moment wind + f_{brkp} = (51,849 ft-lb) + (2.0) [(1,211 lb) (3.9 ft)]

= 61,295 ft-lb ÷ 1,000 lb/kip = 61.3 ft-kip

Ultimate Moment wind + f_{brkp} + debris = 61.3 ft-kip + (2.0) (3.5 kip) (3.9 ft)

= 88,6 ft-kip

Maximum Factored Moment = 88.6 ft-kip = 1,063 in kip

Refer to Table D-3, conservatively assume flood load factor of 2.0 for all axial loads

Maximum factored Axial Compression = (2.0) (30.0 kip) = 60.0 kip

Maximum factored Axial Tension = (2.0) (12.5 kip) = 25.0 kip

Based on a chart published by the Concrete Reinforced Steel Institute (CRSI), the maximum allowable moment for the column = 1,092 in kip for 0 axial load and 1,407 in kip for 102 kip axial load; therefore, the column is adequate.

Check shear strength:

Critical Shear = wind + F_{dyn} + F_i

Ultimate Shear = V_u = [(14 ft) (0.868 kip) (1/3)] (1.6) + [(1.3 kip)+(3.5 kip] (2.0)

= V_u = 16.0 kip

As the maximum unit tension stress is only 25.0 kip/16 in x 16 in = 0.098 kip/in² and the maximum axial compression stress is only 60.0 kip/16 in x 16 in = 0.234 kip/in², we can conservatively treat the column as a flexural member or beam. The allowable shear of the concrete section then, per ACI-318-02 11.3.1.1, 11.3.1.3, and 11.5.5.1 with minimum shear reinforcing (tie/stirrup), would be as follows.

Allowable Shear = V_c = (0.75) (16 in-2.5 in) (16 in) (2) (4,000 psi)$^{1/2}$ (1/1,000) = 20.5 kip

The shear strength of the column is adequate with minimum shear reinforcement.

The minimum area of shear, A_v, per ACI 318-02, 11.5.5.3 would be:

A_v = (50)(width of member)(spacing of reinforcing)/yield strength of reinforcing

= (50)(16 in)(16 in)/(60,000 psi) = 0.21 in²

2 pieces of #4 bar A_s = (2)(0.2) = 0.40 in²

Use of #4 bar for column ties (shear reinforcement) is adequate.

Check spacing per ACI -318-02, 7.10.4

16 diameter of vertical reinforcing bar = (16)(1 in) = 16 in

48 diameter of column tie bar = (48)(1/2 in) = 24 in

Least horizontal dimension = 16 in

Therefore, #4 ties at 16 inches on center are adequate.

The column design is adequate.

Grade Beam Design

The size of the grade beam was configured to provide adequate bearing area, resistance to uplift, a reasonable measure of protection from damaging scour, and to provide a factor of redundancy and reserve strength should the foundation be undermined. A grade beam 36 inches wide and 48 inches deep was selected.

Maximum Bearing Pressure = 1,163 psf = 1.2 ksf (kip/square foot)

Assume a combined load factor of 2.0 (for flood)

Check Shear strength:

Maximum Factored Uniform Bearing Pressure = w_u = (2.0)(3.0 ft)(1.2 ksf) = 7.2 kip/ft

Maximum Factored Shear = V_u = (7.2 kip/ft)(14 ft/2) = 50.0 kip

Allowable Shear without minimum shear reinforcing (stirrups) = $V_c/2$

$V_c/2$ = (0.75)(36 in)(48 - 3.5 in)(63 psi)(1/1,000) = 75.7 kip

Use nominal #4 two leg stirrups at 24 inches on center

Check flexural strength: Assume simple span condition

Maximum Factored Moment = M_u = (7.2 kip/ft)(14 ft)² (1/8) = 176 ft-kip

D FOUNDATION ANALYSIS AND DESIGN EXAMPLES

Concrete strength = 4,000 psi

Reinforcement grade = 60,000 psi

Try (4) #6 reinforcing bar continuous top and bottom

A_s = (4)(0.44 in²) = 1.76 in² top or bottom

total A_s = (2)(1.76) = 3.52 in²

Reinforcement ratio (ρ)

ρ = A_s ÷ [(section width)(section depth- clear cover- ½ bar diameter)]

ρ = A_s ÷ [(b_w)(d)]

ρ = (1.76 in²) ÷ [(36)(48 - 3 - 0.375)]

ρ = 0.01096

One method of calculating the moment strength of a rectangular beam, for a given section and reinforcement, is illustrated in the 2002 edition of the *CRSI Design Handbook*. Referencing page 5-3 of the handbook, the formula for calculating the moment strength can be written as follows:

ΦM_n = (Φ)[((A_s)(f_y)(d)) − (((A_s)(f_y)) ÷ ((0.85)(f'c)(width of member)(2)))]

ΦM_n = (0.9)[((1.76)(60,000)(44.63)) − (((1.76)(60,000)) ÷ ((0.85)(4,000)(36)(2)))]

= 4,241,634 in lb ÷ 12,000 = 353 ft-kip

Φ = 0.9

Area of reinforcing steel (A_s) minimum flexural

= (0.0033)(24)(44.63)

= 3.6 in²

or

(1.33)(A_s required by analysis) = ΦM_n is much greater than M_u

Area of steel for shrinkage and temperature required = (0.0018)(48)(36) = 3.1 in²

Total area of steel provided = (8) #6 = (8)(0.44) = 3.52 in² adequate

Therefore, the grade beam design is adequate, use (4) #6 reinforcing bar continuous on the top and bottom with #4 stirrups at 24-inch spacing.

RECOMMENDED RESIDENTIAL CONSTRUCTION
FOR COASTAL AREAS

Building on Strong and Safe Foundations

E. Cost Estimating

The cost data provided in Appendix E were developed in 2006 for the First Edition of FEMA 550 for select communities along the Gulf of Mexico. The applicability or accuracy of the data to other coastal areas has not been investigated. Although *relative* costs between foundation systems may apply in other coastal regions, users of the manual should verify current *actual* costs for any given location.

E COST ESTIMATING

Breakdown of Foundation Costs			
Foundation Type	Average Foundation Costs ($)	Elevation Above Grade	Unit Costs per Square Foot (sf)
Open			
Case A: Braced Timber Pile	13,536	0 to 5	11
	17,554	5 to 10	15
	22,720	10 to 15	19
	Not Evaluated	Above 15	
Case B: Steel Pipe Pile with Concrete Column and Grade Beam	32,500	0 to 5	27
	36,024	5 to 10	30
	37,500	10 to 15	31
	Not Evaluated	Above 15	
Case C: Timber Pile with Concrete Column and Grade Beam	31,700	0 to 5	26
	36,288	5 to 10	30
	37,900	10 to 15	32
	Not Evaluated	Above 15	
Case D: Concrete Column and Grade Beam	13,500	0 to 5	11
	16,860	5 to 10	14
	18,500	10 to 15	15
	Not Evaluated	Above 15	
Case G: Concrete Column and Grade Beam with Integral Slab	18,000	0 to 5	15
	21,847	5 to 10	18
	24,000	10 to 15	20
	Not Evaluated	Above 15	
Closed			
Case E: Reinforced Masonry - Crawlspace	12,254	0 to 4	10
	14,000	4 to 8	12
Case F: Reinforced Masonry - Stem Wall	12,458	0 to 4	10

NOTES

1. This rough order of magnitude (ROM) cost estimate is based upon May 2006 figures for concrete, labor, equipment, and materials. Variations due to labor/equipment/materials shortages are anticipated and should be taken into account when using these costs.

2. Costs presented herein should not be construed to represent actual costs to the homebuilder, but should be utilized as an order of magnitude estimate only.

3. Pile driving mobilization/demobilization can be reduced if several homes are constructed at the same time in the same area, thereby realizing an economy of scale.

COST ESTIMATING E

4. Costs presented are based upon the general designs in this document. A 1,200-sf footprint for a single-story home at an assumed 130-mph wind speed, elevated to the average height for that foundation, is the basis for the estimates. When differences in elements of construction occur, such as number of piles or amount of concrete, an alternate cost is presented. The cost estimate presented represents the conservative approach to the designs in this document. If value engineering, different materials, or a more cost-effective design are implemented, these costs may be reduced.

5. Costs presented herein include applicable taxes, contractor general and home office overhead, profit, and other sub-tier contract costs.

6. Concrete costs, unless otherwise noted, include bracing, reinforcing, formwork, finishing (if necessary), and mobilization and demobilization of the contractor. Due to the anticipated shortage of skilled labor for concrete, variability in this area should be anticipated.

7. Costs experienced by the builder or contractor will be dependent upon contract agreements, local price variations in labor, material, equipment, and availability.

8. Costs for steel are highly variable and dependent upon supply. Variability in costs for steel should be anticipated. Costs for steel include materials and labor for installation.

9. Costs for block in closed foundations is based upon standard natural gray medium weight masonry block walls, including blocks, mortar, typical reinforcing, normal waste, and walls constructed with 8" x 8" x 16" blocks laid in running bond. Add for grouting cores poured by hand to 4-foot heights.

E COST ESTIMATING

<td colspan="9" align="center">**Case A: Braced Timber Pile**</td>
	Unit of Measure	Material	Labor	Equip.	Subtotal	Number of Piles	Length of Piles Driven	Total lf to drive	Subtotal
Site Prep	ls				500.00	—	—	—	$500
Minimum job charge for Driving	ls	—	—	—	5,000.00	—	—	—	$5,000
<td colspan="9" align="center">**Number of Piles: 60**</td>
Over 30' to 40' (800 lf per day)	lf	3.3	1.7	0.9	5.90	60	30	1,800	$10,620
Bolts and Miscellaneous	Per Column				15.00	60			$900
Wood Pile Connection to House	Per Pile				55.00	60			$3,300
Galvanized Bracing Rod and Turnbuckles	Per Pile				40.00	60			$2,400
Total for Piles									$22,720
<td colspan="9" align="center">**Number of Piles: 42**</td>
Over 30' to 40' (800 lf per day)	lf	3.3	1.7	0.9	5.90	42	30	1260	$7,434
Bolts and Miscellaneous	Per Column				15.00	42			$630
Wood Pile Connection to House	Per Pile				55.00	42			$2,310
Galvanized Bracing Rod and Turnbuckles	Per Pile				40.00	42			$1,680
Total for Piles									$17,554
<td colspan="9" align="center">**Number of Piles: 28**</td>
Over 30' to 40' (800 lf per day)	lf	3.3	1.7	0.9	5.90	28	30	840	$4,956
Bolts and Miscellaneous	Per Column				15.00	28			$420
Wood Pile Connection to House	Per Pile				55.00	28			$1,540
Galvanized Bracing Rod and Turnbuckles	Per Pile				40.00	28			$1,120
Total for Piles									$13,536

ls = lump sum   lf = linear foot

COST ESTIMATING

Case B: Steel Pipe Pile with Concrete Column and Grade Beam									
Site Prep	ls				500.00	—	—	—	$500
Minimum job charge for Driving	ls	—	—	—	5,000.00	—	—	—	$5,000
Steel Piles Driven									
	Unit of Measure	Material	Labor	Equip.	Subtotal	Number of Piles	Length of Piles	Total lf to drive	Subtotal
Steel Piles Driven	lf	11	3.1	0.9	15.00	28	30	840	$12,600
Bolts and Miscellaneous	Per Column				25.00	28			$700
Total for Piles									$13,300
Concrete Grade Beam									
	Unit of Measure	Material	Labor	Equip.	Subtotal	Number of cy	Equipment Charge		Subtotal
Soil Excavation, Medium material, 75 cy per hour (57 m³/hr)	cy	—	0.54	1.29	1.83	55	500		$601
Grade beams	cy	225	22	7.5	254.50	32			$8,144
Steel	ea				100.00	32			$3,200
Total for Grade Beams									$11,945
Concrete Columns @ 10 feet									
	Unit of Measure	Material	Labor	Equip.	Subtotal	Number of cy/Columns	Number of Columns		Subtotal
18" (46 cm) square or round columns	cy	225	35	7.5	267.50	0.74	12		$2,375
Steel	Column				150.00		12		$1,800
Anchors	Column				49.55		12		$595
Angles	Column				42.45		12		$509
Subtotal for Concrete Columns									$5,279
Total for Case B									$36,024

ls = lump sum   lf = linear foot   cy = cubic yard   ea = each

RECOMMENDED RESIDENTIAL CONSTRUCTION FOR COASTAL AREAS

E COST ESTIMATING

Case C: Timber Pile with Concrete Column and Grade Beam

	Unit of Measure	Material	Labor	Equip.	Subtotal	Number of Piles	Length of Piles	Total lf to Drive
Site Prep	ls				500.00	—	—	—
Minimum job charge for Driving	ls	—	—	—	5,000.00	—	—	—
Timber Piles Driven								
Number of Wooden Piles: 42								
Over 30' to 40' (800 lf per day)	lf	3.3	1.7	0.9	5.90	42	30	1,260
Bolts and Miscellaneous	Per Column				15.00	42		
Total for Piles								

Concrete Grade Beam

	Unit of Measure	Material	Labor	Equip.	Subtotal	Number of cy	Equipment Charge	
Soil Excavation, Medium material, 75 cy per hour (57 m³/hr)	cy	—	0.54	1.29	1.83	55	500	
Grade beams	cy	225	22	7.5	254.50	32		
Steel	ea				100.00	32		
Total for Grade Beams								

Concrete Columns including Pile Caps @ 10 feet

	Unit of Measure	Material	Labor	Equip.	Subtotal	Number of cy/ Columns	Number of Columns	
18" (46 cm) square or round columns	cy	225	35	7.5	267.50	0.74	12	
Steel	Column				150.00		12	
Anchors	Column				49.55		12	
Angles	Column				42.45			
Subtotal for Concrete Columns								
Grand Total for Foundation Shown								

ls = lump sum   lf = linear foot   cy = cubic yard   ea = each

COST ESTIMATING E

	Case D: Concrete Column and Grade Beam							
	Unit of Measure	Material	Labor	Equip.	Subtotal	Number of cy	Equipment Charge	Subtotal
Soil Excavation, Medium material, 75 cy per hour (57 m³/hr)	cy	—	0.54	1.29	1.83	50	500	$592
Grade beams	cy	225	22	7.5	254.50	31		$7,890
Steel	ea				100	31		$3,100
Total for Grade Beams								$11,582
Concrete Columns @ 10 feet Elevation								
	Unit of Measure	Material	Labor	Equip.	Subtotal	Number of cy per Column	Number of Columns	Subtotal
18" (46 cm) square or round columns	cy	225	35	7.5	267.50	0.74	12	$2,375
Steel	Column				150.00		12	$1,800
Anchors	Column				49.55		12	$595
Angles	Column				42.45		12	$509
Subtotal for Concrete Columns								$5,279
Grand Total for Foundation Shown								$16,861

cy = cubic yard   ea = each

RECOMMENDED RESIDENTIAL CONSTRUCTION FOR COASTAL AREAS   E-7

E COST ESTIMATING

Case G: Concrete Column and Grade Beam with Integral Slab								
	Unit of Measure	Material	Labor	Equip.	Subtotal	Number of cy	Equipment Charge	Subtotal
Soil Excavation, Medium material, 75 cy per hour (57 m³/hr)	cy	—	0.54	1.29	1.83	75	500	$637
Interior Concrete	cy	225	35	7.50	267.50	15		$4,013
Grade Beams	cy	225	35	7.50	267.50	31		$8,293
Steel	ea				100.00	31		$3,100
WWF	ea				35.00	15		$525
Total for Grade Beams								**$16,568**
Concrete Columns @ 10 feet Elevation								
	Unit of Measure	Material	Labor	Equip.	Subtotal	Number of cy per Column	Number of Columns	Subtotal
18" (46 cm) square or round columns	cy	225	35	7.50	267.50	0.74	12	$2,375
Steel	Column				150		12	$1,800
Anchors	Column				49.55		12	$595
Angles	Column				42.45		12	$509
Subtotal for Concrete Columns								**$5,279**
Grand Total for Foundation Shown								**$21,847**

cy = cubic yard   ea = each   WWF = welded wire fabric

COST ESTIMATING E

Case E: Reinforced Masonry – Crawlspace							
Crawlspace							
	Unit of Measure	Material	Labor	Equip.	Subtotal	Number of cy	Equipment Charge	Subtotal
Soil Excavation, Medium material, 75 cy per hour (57 m³/hr)	cy	—	0.68	1.43	2.11	24	500	$551
Footings	cy	225	22	7.5	254.50	20		$5,090
Steel	cy				100.00	20		$2,000
Total for Footings (not including steel)								$7,641
Concrete Columns @ 4 feet							
	Unit of Measure	Material	Labor	Equip.	Subtotal	Number of cy per Column	Number of Columns	Subtotal
16" (46 cm) square or round columns piers	cy	225	35	7.5	267.50	0.40	6	$642
Steel	Column				150.00		6	$900
Anchors	Column				49.55		6	$297
Angles	Column				42.45		6	$255
Subtotal for Concrete Columns								$2,094
Concrete Walls							
	Unit of Measure	Material	Labor	Equip.	Subtotal	Number of sf		Subtotal
Concrete Block Walls	sf	5.23	4.53	1	10.76	420		$4,519
Grand Total for Foundation Shown (Footings + Concrete Columns + Concrete Walls)								$14,254

cy = cubic yard   sf = square foot

E COST ESTIMATING

	Unit of Measure	Material	Labor	Equip.	Subtotal	Number of cy	Equipment Charge	Subtotal
	Case F: Reinforced Masonry – Stem Wall
Soil Excavation, Medium material, 75 cy per hour (57 m³/hr)	cy	—	0.54	1.29	1.83	34	500	$562
Footings	cy	225	22	7.5	254.50	20		$5,090
Steel	cy				100.00	20		$2,000
Bracing	ls							$1,500
Backfill	cy	4	0.6	1.45	6.05	130		$787
Total for Stem Wall Footings								$9,152
	Concrete Walls
	Unit of Measure	Material	Labor	Equip.	Subtotal	Number of sf		Subtotal
Concrete Block Walls	sf	5.23	4.53	1	10.76	420		$4,519
Grand Total for Foundation Shown								$13,671

cy = cubic yard   ls = lump sum   sf = square foot

RECOMMENDED RESIDENTIAL CONSTRUCTION
FOR COASTAL AREAS

Building on Strong and Safe Foundations

F. Pertinent Coastal Construction Information

FEMA 499 Fact Sheet No.	Title
1	Coastal Building Successes and Failures
2	Summary of Coastal Construction Requirements and Recommendations
4	Lowest Floor Elevation
5	V-Zone Design and Construction Certification
6	How Do Siting and Design Decisions Affect the Owner's Costs?
7	Selecting a Lot and Siting the Building
8	Coastal Building Materials
9	Moisture Barrier Systems
11	Foundations in Coastal Areas

F PERTINENT COASTAL CONSTRUCTION INFORMATION

12	Pile Installation
13	Wood-Pile-to-Beam Connections
14	Reinforced Masonry Pier Construction
15	Foundation Walls
16	Masonry Details
26	Shutter Alternatives
27	Enclosures and Breakaway Walls
29	Protecting Utilities

All of the FEMA 499 Fact Sheets listed above can be viewed, downloaded, or printed. Go to http://www.fema.gov/library/viewRecord.do?id=1570 and click on the Resource File for the Fact Sheet(s) of interest.

FEMA P-757 Recovery Advisory

Erosion, Scour, and Foundation Design

The above Recovery Advisory can be viewed, downloaded, or printed. Go to http://www.fema.gov/library/viewRecord.do?id=3577 and click on Appendix D, pages 20-23.

G. FEMA Publications and Additional References

The American Concrete Institute (ACI). 2005. *Building Code Requirements for Structural Concrete and Commentary*, ACI 318.

American Forest & Paper Association's (AF&PA) American Wood Council (AWC). 2005. *National Design Specification for Wood Construction*. ANSI/AF&PA NDS 2005.

American Forest & Paper Association. 2008. *Wood Frame Construction Manual (WFCM) for One- and Two-Family Dwellings*.

American Society for Testing and Materials (ASTM). 2000. *Standard Test Method for High-Strain Dynamic Testing of Piles*, ASTM D 4945. November 2000.

American Society of Civil Engineers (ASCE) 7-05. 2005. *Minimum Design Loads for Buildings and Other Structures*, ASCE 7-05. ISBN: 0784408092.

ASCE 24-05. 2006. *Flood Resistant Design and Construction*. ISBN: 0784408181.

American Wood Preservers' Association (AWPA). 1991. *Care of Pressure-Treated Wood Products*.

American Wood Preservers' Association. 2008. *AWPA M4-08 Standard for the Care of Preservative-Treated Wood Products*.

Concrete Reinforcing Steel Institute (CRSI). 2002. *CRSI Design Handbook*.

G FEMA PUBLICATIONS AND ADDITIONAL REFERENCES

FEMA. 1984. *Elevated Residential Structures*, FEMA 54. March 1984.

FEMA. 1995. *Guide to Flood Maps: How to Use a Flood Map to Determine Flood Risk for a Property*, FEMA 258. May 1995.

FEMA. 1996. *Corrosion Protection for Metal Connectors in Coastal Areas for Structures Located in Special Flood Hazard Areas*, FEMA Technical Bulletin 8-96. August 1, 1996.

FEMA. 2000. *Coastal Construction Manual*, FEMA 55. June 2000.

FEMA. 2004. *Design Guide for Improving School Safety in Earthquakes, Floods, and High Winds*, FEMA 424. January 2004.

FEMA. 2005. *Mitigation Assessment Team Report: Hurricane Charley in Florida*, FEMA 488. April 2005.

FEMA. 2005. *Mitigation Assessment Team Report: Hurricane Ivan in Alabama and Florida*, FEMA 489. August 2005.

FEMA. 2005. *Home Builders' Guide to Coastal Construction Technical Fact Sheets*, FEMA 499. August 2005.

FEMA. 2006. *Mitigation Assessment Team Report: Hurricane Katrina in the Gulf Coast*, FEMA 549. July 2006.

FEMA. 2006. *Flood Insurance Manual.* http://www.fema.gov/business/nfip/manual.shtm. May 2006.

FEMA. 2008. NFIP Technical Bulletin 2, *Flood Damage-Resistant Materials Requirements for Buildings Located in Special Flood Hazard Areas in accordance with the National Flood Insurance Program.* August 2008

FEMA. 2009. Hurricane Ike MAT *Recovery Advisories: Enclosures and Breakaway Walls* and *Erosion, Scour, and Foundation Design.* FEMA P-757. April 2009.

International Building Code (IBC). *2009 International Building Code.* 2009.

International Code Council (ICC). *Standard for Residential Construction in High Wind Regions* (ICC-600). November 2006.

International Residential Code (IRC). *2009 International Residential Code for One- and Two-Family Dwellings.* 2009.

Mississippi Governor's Rebuilding Commission on Recovery, Rebuilding and Renewal. 2005. *A Pattern Book for Gulf Coast Neighborhoods.* November 2005.

National Fire Protection Association (NFPA). 2003. *NFPA 5000®: Building Construction and Safety Code®*, 2003 Edition.

RECOMMENDED RESIDENTIAL CONSTRUCTION
FOR COASTAL AREAS
Building on Strong and Safe Foundations

H. Glossary

3-second peak gust – The wind speed averaging time used in ASCE 7 and the IBC.

Allowable Stress Design (ASD) – A method of proportioning structural members such that elastically computed stresses produced in the members by nominal loads do not exceed specified allowable stresses (also called working stress design).

A Zone – A zones are the areas not listed as V zones, but also identified on a Flood Insurance Rate Map (FIRM) as being subject to inundation during a 100-year flood. The associated flood elevation has a 1-percent chance of being equaled or exceeded in any given year. There are several categories of A zones that may be identified on a FIRM with one of the following designations: AO, AH, A1-30, AE, and unnumbered A zones.

Base flood – A flooding having a 1-percent chance of being equaled or exceeded in any given year; also known as the 100-year flood.

Base Flood Elevation (BFE) – Elevation of the 1-percent flood. This elevation is the basis of the insurance and floodplain management requirements of the National Flood Insurance Program (NFIP).

Coastal A zone – The portion of the Special Flood Hazard Area (SFHA) landward of a V zone or landward of an open coast without mapped V zones (e.g., the shorelines of the Great Lakes), in which the principal sources of flooding are astronomical tides, storm surges, seiches, or

H GLOSSARY

tsunamis, not riverine sources. Like the flood forces in V zones, those in Coastal A zones are highly correlated with coastal winds or coastal seismic activity. Coastal A zones may therefore be subject to wave effects, velocity flows, erosion, scour, or combinations of these forces. During base flood conditions, the potential for breaking wave heights between 1.5 feet and 3.0 feet will exist. Coastal A zones are not shown on present day FIRMs or mentioned in a community's Flood Insurance Study (FIS) Report.

Crawlspace foundation – Crawlspace foundations are typically low masonry perimeter walls with interior piers supporting a wood floor system. These foundations are usually supported by shallow footings and are prone to failure caused by erosion or scour.

Design flood – The design flood is often, but not always equal to the base flood for areas identified as SFHAs on a community's FIRM.

Design Flood Elevation (DFE) – The elevation of the design flood, including wave height relative to the datum specified on a community's Flood Hazard Map.

Design professional – A State licensed architect or engineer.

Erosion – Process by which floodwaters lower the ground surface in an area by removing upper layers of soil.

Exposure Category B – A wind exposure identified in ASCE 7 and the International Building Code (IBC) as urban and suburban areas, wooded areas, or other terrain with numerous closely spaced obstructions having the size of single-family dwellings or larger.

Exposure Category C – A wind exposure identified in ASCE 7 and the IBC as open terrain with scattered obstructions having heights generally less than 30 feet (9.1 meters). This category includes flat open country, grasslands, and all water surfaces in hurricane-prone regions.

Fixity – Resistance to flotation; stable or immovable.

Flood Insurance Rate Map (FIRM) – An official map of a community, on which FEMA has delineated both the SFHA and the risk premium zones applicable to the community. The map shows the extent of the base floodplain and may also display the extent of the floodway and BFEs.

Freeboard – The height added to place a structure above the base flood to reduce the potential for flooding. The increased elevation of a building above the minimum design flood level to provide additional protection for flood levels higher than the 1-percent chance flood level and to compensate for inherent inaccuracies in flood hazard mapping.

Hydrodynamic forces – The amount of pressure exerted by moving floodwaters on an object, such as a structure. Among these loads are positive frontal pressure against the structure, drag forces along the sides, and suction forces on the downstream side.

Hydrostatic forces – The amount of lateral pressure exerted by standing or slowly moving floodwaters on a horizontal or vertical surface, such as a wall or a floor slab. The water pressure increases with the square of the water depth.

Leeward – The side away, or sheltered, from the wind.

Limit of Moderate Wave Action (LiMWA) – The landward extent of coastal areas designated Zone AE where waves higher than 1.5 feet can exist during a design flood (also known as the Coastal A zone).

National Flood Insurance Program (NFIP) – The NFIP is a Federal program enabling property owners in participating communities to purchase insurance as a protection against flood losses in exchange for State and community floodplain management regulations that reduce future flood damages. Participation in the NFIP is based on an agreement between communities and the Federal Government. If a community adopts and enforces floodplain management regulations to reduce future flood risks to new construction in floodplains, the Federal Government will make flood insurance available within the community as a financial protection against flood losses. This insurance is designed to provide an insurance alternative to disaster assistance to reduce the escalating costs of repairing damage to buildings and their contents caused by floods. The program was created by Congress in 1968 with the passage of the National Flood Insurance Act of 1968.

National Geodetic Vertical Datum 1929 (NGVD 1929) – A vertical elevation baseline determined in 1929 as a national standard. Used as the standard for FIRMs until 2000.

North American Vertical Datum 1988 (NAVD 1988) – A vertical elevation baseline determined in 1988 as a more accurate national standard. The current vertical elevation standard for new FIRMs.

Scour – Erosion by moving water in discrete locations, often as a result of water impacting foundation elements.

Shore-normal – Perpendicular to the shoreline.

Slab-on-grade foundation – Type of foundation in which the lowest floor of the house is formed by a concrete slab that sits directly on the ground.

Slug – A unit of mass in the English foot-pound-second system. One slug is the mass accelerated at 1 foot per second (fps) by a force of 1 pound (lb). Since the acceleration of gravity (g) in English units is 32.174 fps, the slug is equal to 32.174 pounds (14.593 kilograms).

Special Flood Hazard Area (SFHA) – Portion of the floodplain subject to inundation by the base flood.

Stem wall foundation – A type of foundation that uses masonry block and reinforced with steel and concrete. The wall is constructed on a concrete footing, back-filled with dirt, compacted, and the slab is then poured on top.

H GLOSSARY

Strength Design – A method of proportioning structural members such that the computed forces produced in the members by the factored loads do not exceed the member design strength (also called load and resistance factor design).

V zones – V zones are areas identified on FIRMs as zones VE, V1-30, or V. These areas, also known as Coastal High Hazard Areas, are areas along the coast that have a 1 percent or greater annual chance of flooding from storm surge and waves greater than 3 feet in height, as well as being subject to significant wind forces.

Wave trough – The lowest part of the wave between crests.

Windward – The side facing the wind.

RECOMMENDED RESIDENTIAL CONSTRUCTION
FOR COASTAL AREAS

Building on Strong and Safe Foundations

I. Abbreviations and Acronyms

ABFE	Advisory Base Flood Elevation
ACI	American Concrete Institute
AF&PA	American Forest & Paper Association
AISI	American Iron and Steel Institute
AL	Alabama
ANSI	American National Standards Institute
ASCE	American Society of Civil Engineers
ASD	allowable stress design
ASTM	American Society for Testing and Materials

ABBREVIATIONS AND ACRONYMS

AWC American Wood Council

AWPA American Wood Preservers' Association

B

BFE Base Flood Elevation

C

C&C components and cladding

CMU concrete masonry unit

CRSI Concrete Reinforced Steel Institute

cy cubic yard

D

DFE Design Flood Elevation

E

ea each

F

FEMA Federal Emergency Management Agency

FIRM Flood Insurance Rate Map

FIS Flood Insurance Study

FL Florida

ABBREVIATIONS AND ACRONYMS

fps feet per second

FRP fiber reinforced polyester

ft feet

G

g gravity

IBC International Building Code

ICC International Code Council

IRC International Residential Code

kip 1,000 pounds

ksi kips per square inch

L

lb pound

LA Louisiana

lb/sq pounds per square foot

lf linear foot

LiMWA Limit of Moderate Wave Action

ls lump sum

ABBREVIATIONS AND ACRONYMS

M

MAT	Mitigation Assessment Team
mph	miles per hour
m/s	meters per second
MS	Mississippi
MWFRS	main wind force resisting system

N

NAVD	North American Vertical Datum
NDS	National Design Specification
NFIP	National Flood Insurance Program
NFPA	National Fire Protection Association
NGS	National Geodetic Survey
NGVD	National Geodetic Vertical Datum
NR	Not Recommended
NRCS	National Resource Conservation Service

O

o.c.	on center

P

PCA	Portland Cement Association

ABBREVIATIONS AND ACRONYMS

pcf	pounds per cubic foot
p/lf	pounds per linear foot
psf	pounds per square foot
psi	pounds per square inch

RA	Recovery Advisory
RISA	Rapid Interactive Structure Analysis
ROM	rough order of magnitude

SBC	Standard Building Code
SCS	Soil Conservation Service
SEI	Structural Engineering Institute
SFHA	Special Flood Hazard Area
SLOSH	Sea, Lake, and Overland Surges from Hurricanes
sq	square foot
SWL	stillwater level

T

TBTO	tributylin oxide
TMS	The Masonry Society
TX	Texas

ABBREVIATIONS AND ACRONYMS

UDA	Urban Design Associates
u.n.o.	unless noted otherwise
USGS	United States Geological Survey

WCFM	Wood Frame Construction Manual for One- and Two-Family Dwellings
WWF	welded wire fabric

www.ingramcontent.com/pod-product-compliance
Lightning Source LLC
Chambersburg PA
CBHW082116230426
43671CB00015B/2718